26. Colloquium der Gesellschaft für Biologische Chemie
10.-12. April 1975 in Mosbach/Baden

Molecular Basis of Motility

Edited by
L. M. G. Heilmeyer Jr., J. C. Rüegg and Th. Wieland

With 103 Figures

Springer-Verlag
Berlin Heidelberg New York 1976

ISBN 3-540-07576-3 Springer-Verlag Berlin   Heidelberg   New York
ISBN 0-387-07576-3 Springer-Verlag New York   Heidelberg   Berlin

Library of Congress Cataloging in Publication Data. Gesellschaft für Biologische Chemie. Molecular basis of motility. Bibliography:
p. Includes index. 1. Muscle-Motility-Congresses. 2. Muscle contraction-Congresses. I. Heilmeyer, L., 1937—. II. Rüegg, Johann C.,
1930—. III. Wieland, Theodor. IV Title. QP321.G38.1976.591.1′852.76-2482.

© by Springer-Verlag Berlin · Heidelberg 1976.

Printed in Germany.

Offsetprinting and bookbinding: Brühlsche Universitätsdruckerei, Gießen.

Contents

Contributors

Alemà, S.
Laboratory of Cell Biology, C.N.R., Via Romagnosi 18A, Rome, Italy

Bagshaw, Clive R.
Department of Biochemistry, School of Medicine, University of Pennsylvania, Philadelphia, PA 19174, USA

Barrington Leigh, J.
European Molecular Biology Laboratory, Außenstation DESY, 2000 Hamburg, FRG

Becker, J.-U.
Department of Biochemistry, University of Washington, Seattle, WA 98195, USA

Blum, H.E.
Department of Biochemistry, University of Washington, Seattle, WA 98195, USA

Bryan, Joseph
Biology Department, Leidy Laboratory, University of Pennsylvania, Philadelphia, PA 19174, USA

Byers, B.
Department of Biochemistry, University of Washington, Seattle, WA 98195, USA

Calissano, P.
Laboratory of Cell Biology, C.N.R., Via Romagnosi 18A, Rome, Italy

Chen, J.S.
Laboratory of Cell Biology, C.N.R., Via Romagnosi 18A, Rome, Italy

Cole, H.A.
Department of Biochemistry, The University of Birmingham, Birmingham, Great Britain

Collins, J.H.
Department of Muscle Research, Harvard Medical School, Boston, MA 02114, USA

Eccleston, J.F.
Department of Biochemistry, University of Bristol, Medical School, Bristol, BS8 1TD, Great Britain

Fischer, Edmond
Department of Biochemistry, University of Washington, Seattle, WA 98195, USA

Frearson, N.
Department of Biochemistry, The University of Birmingham, Birmingham, Great Britain

Geeves, Michael A.
Department of Biochemistry, Medical School, University of Bristol, Bristol, BS8 1TD, Great Britain

Gergely, J.
Department of Muscle Research, Boston Biomedical Research Institute, Harvard Medical School, Boston, MA 02114, USA

Goody, R.S. Max-Planck-Institut für medizinische Forschung, Jahnstraße 29, 6900 Heidelberg, FRG

Hasselbach, W. Max-Planck-Institut für medizinische Forschung, Abt. Physiologie, Jahnstraße 29, 6900 Heidelberg, FRG

Heilmeyer, Jr. Institut für Physiologische Chemie der Universität,
Ludwig, M.G. Universitätsstraße 150, 4630 Bochum-Querenburg, FRG

Heizmann, C. Department of Biochemistry, University of Washington, Seattle, WA 98195, USA

Holmes, Kenneth C. Max-Planck-Institut für medizinische Forschung, Jahnstraße 29, 6900 Heidelberg, FRG

Huxley, H.E. MRC Laboratory of Molecular Biology, Hills Road, Cambridge CB2 2QH, Great Britain

Kendrick-Jones, MRC Laboratory of Molecular Biology, Hills Road,
John Cambridge CB2 2QH, Great Britain

Kerrick, G.W. Department of Biochemistry, University of Washington, Seattle, WA 98195, USA

Leavis, P. Department of Muscle Research, Boston Biomedical Research Institute, Harvard Medical School, Boston, MA 02114, USA

Lehky, P. Department of Biochemistry, University of Washington, Seattle, WA 98195, USA

Lehrer, S.S. Department of Muscle Research, Boston Biomedical Research Institute, Harvard Medical School, Boston, MA 02114, USA

Levi, A. Laboratory of Cell Biology, C.N.R., Via Romagnosi 18A, Rome, Italy

Levi-Montalcini, Laboratory of Cell Biology, C.N.R., Via Romagnosi
R. 18A, Rome, Italy

Lüscher, E.F. Theodor Kocher Institut, Universität Bern, Freie Straße 1, CH-3000 Bern 9

Malencik, D.A. Department of Biochemistry, University of Washington, Seattle, WA 98195, USA

Mannherz, H.G. Max-Planck-Institut für medizinische Forschung, Jahnstraße 29, 6900 Heidelberg, FRG

Moir, A.J.G. Department of Biochemistry, The University of Birmingham, Birmingham, Great Britain

Morgan, M. Department of Biochemistry, The University of Birmingham, Birmingham, Great Britain

Mrwa, Ulrike II. Physiologisches Institut der Universität Heidelberg, Im Neuenheimer Feld 326, 6900 Heidelberg, FRG

Nagle, Barbara W.   Biology Department, Leidy Laboratory, University of Pennsylvania, Philadelphia, PA 19174, USA

Nagy, B.   Department of Muscle Research, Boston Biomedical Research Institute, Harvard Medical School, Boston, MA 02114, USA

Perry, S.V.   Department of Biochemistry, The University of Birmingham, Birmingham, Great Britain

Pires, E.   Department of Biochemistry, The University of Birmingham, Birmingham, Great Britain

Pocinwong, S.   Department of Biochemistry, University of Washington, Seattle, WA 98195, USA

Podolsky, Richard J.   National Institute of Arthritis, Metabolism, and Digestive Diseases, National Institutes of Health, Bethesda, MD 20014, USA

Potter, J.D.   Department of Muscle Research, Boston Biomedical Research Institute, Harvard Medical School, Boston, MA 02114, USA

Rosenbaum, G.   European Molecular Biology Laboratory, Außenstation DESY, 2000 Hamburg, FRG

Rüegg, J.C.   II. Physiologisches Institut, Universität Heidelberg, Im Neuenheimer Feld 326, 6900 Heidelberg, FRG

Seidel, J.C.   Department of Muscle Research, Boston Biomedical Research Institute, Harvard Medical School, Boston, MA 02114, USA

Trentham, David R. Department of Biochemistry, Medical School, University of Bristol, Bristol, BS8 1TD, Great Britain

Wieland, Th.   Max-Planck-Institut für medizinische Forschung, Abt. Naturstoff-Chemie, Jahnstraße 29, 6900 Heidelberg, FRG

Wilkie, D.R.   Department of Physiology, University College London, Gower Street, London, WC1E 6BT, Great Britain

Welcome

Herr Vorsitzender der biochemischen Gesellschaft,
Herr Bürgermeister von Mosbach,
Meine Damen und Herren!

Ich möchte Sie herzlich begrüßen und am Symposium über molekulare
Grundlagen der Motilität willkommen heißen. Sie erlauben mir jetzt
sicher, vor allem im Hinblick auf unsere vielen Gäste und der Inter-
nationalität dieses Symposiums wegen, Englisch weiterzufahren, with
a cordial welcome to our guests. For many years I have wished that we
would have a Symposium on muscle in our country and now this wish has
become true, as the German Biochemical Society decided to have this
years' Mosbach Symposium devoted to problems of the molecular basis
of motility. When, about a year ago, Ludwig Heilmeyer and myself sat
together and thought about the meeting, it seemed to us a matter of
course that the Doyen, or the grand old man of German Muscle Physiology,
Hans-Hermann Weber, should open this meeting with an introduction to
muscle research past and present to which he has contributed so much,
especially in the field of chemo mechanical energy conversion and in
the study of the rôle of ATP during contraction. This hope however,
could not be realized because of the unexpected death of Professor
Weber last June. For this reason we are all here now also in memoriam
H.H. Weber who meant so much to all of us. Prof. Hasselbach, who knew
him for at least one decade longer than I, will recall his life and
scientific work a little later in his introduction to this meeting.

Though H.H. Weber cannot be with us, we have at least the consolation
that his daughter Annemarie, who by all standards is one of the most
outstanding musclebiochemists, is able to be here. I hope that she
will happy to see that the muscle research which her father commenced
continues to flourish and that his seed of muscle physiology is growing
well in Germany. We are very grateful indeed that you were able to
come to this Mosbach Symposium.

We are also grateful to all the other invited guest speakers who were
able to come and to contribute to the success of this meeting. As you
all know the success of these meetings relies in no small way on the
high standard of the invited speakers and therefore we decided to in-
vite the top scientists including those from overseas (overseas of
course includes England which is still separated by the channel from
Continental Europe although with the European Molecular Biology Organization, EMBO, now being in Heidelberg, ties to English science will
hopefully get even closer). As you can imagine, as the organizers, we
were concerned whether all would accept - and thank heavens, they did
despite the well-known overload with such meetings. I take this op-
portunity to express my deep gratitude to all our guests and invited
speakers who, by their coming, make this meeting a success. I sincerely
hope that you will have a very enjoyable stay here.

It is a pleasure to thank also the German Biochemical Society represented here by its president, Prof. Helmreich from Würzburg and its
officers, Prof. Auhagen and Prof. Gibian whose efforts in organizing
things, financially and otherwise, made this meeting possible. I would
like to thank also my colleagues in the organizing Committee, Theodor
Wieland and Ludwig Heilmeyer for their unceasing efforts in organizing
the scientific part of this meeting.

Now, I should like to add a few words about the programme. Today we
concentrate on the problem of chemo-mechanical energy conversion and
we focus our attention on these molecular force generators and energy
converters, the myosin cross bridges, on which much interdisciplinary
muscle research converges. There is a structural and chemical approach,
a mechanical and enzyme kinetics approach towards the understanding of
the crossbridge cycle. We hope for a strong interaction between these
various approaches. There is furthermore also the thermodynamic approach to the problem, since after all, even cross-bridges must obey
thermodynamic laws and if they apparently don't something must be wrong
somewhere. I hope that by the end of the day we will all know that all
is well with our cross-bridge cycles. But perhaps we will be told that
we still do not know how chemical energy is transformed and many open
questions will be pointed out. It is my hope that this will stimulate
some of our younger colleagues to enter the field of muscle research
and attempt to solve these problems.

Tomorrow is devoted to regulations of energy conversion. Since the important discovery that energy conversion is controlled by trace calcium,
research has come a long way, especially in the study of regulatory
proteins. The importance of contractile phenomena for many life functions in general can hardly be overemphasized, as will be brought in
the Saturday session on primitive cell motility. In summary, research
on contractile phenomena is now taking a very rapid course.

I am glad that this meeting is taking place here in Mosbach which is
exactly half way between Heidelberg and Würzburg, the two cities where
a hundred years ago Willi Kühne and Adolf Fick initiated the study of
the myosin molecule on the one hand, and of mechano-chemical energy
conversion on the other, studies which, associated with people like
Meyerhof and H.H. Weber, have quite a tradition in this region of
Germany. Besides the merit of lying exactly between Heidelberg and
Würzburg, Mosbach has many other advantages which make it a nice place
for a meeting, advantages which will now surely be pointed out by
Mosbach's Bürgermeister to who I would like to give the word in a
moment.

J.C. Rüegg

Hans Hermann Weber
In Memoriam

Es ist das erste Mal, daß in Deutschland ein Symposium über die kontraktilen Strukturen von Muskeln und Zellen stattfindet. Alle, die auf
diesem Grenzgebiet zwischen Physiologie, Biochemie und Biophysik in
Deutschland arbeiten und gearbeitet haben, sollten sich darüber freuen,
daß ihr Arbeitsgebiet diese Anerkennung gefunden hat.

Diese Freude ist getrübt, weil der Tod uns den Nestor der deutschen
Muskelphysiologie, Hans Hermann Weber, im letzten Jahr genommen hat.

Ich habe ihm noch von den Vorbereitungen zu diesem Symposium erzählen
können. Er hat sich über den Plan gefreut; hat er doch bis zuletzt
die Wissenschaft und ihre Entwicklung mit Interesse verfolgt. Es wäre
ihm eine Freude gewesen, hier zu präsidieren und die Diskussion mit
Fragen zu beleben.

Daß wir H.H. Webers in Mosbach gedenken, scheint mir eine besondere
Fügung. H.H. Weber war der Physiologe unter den Begründern der Mos-
bacher Tagungen, die sich vor 25 Jahren hier zusammenfanden. 1950 be-
durfte es einer großen Anstrengung, nicht zu resignieren. Weber gehörte
zu den Wissenschaftlern, die den Neubeginn wagten und dazu beigetragen
haben, daß Biochemie und Biophysik in Deutschland wieder Anschluß an
die internationale Entwicklung fanden.

Ein Blick auf das Leben Webers zeigt uns die schwierige Situation, in
der sich seine Generation - besonders seine Jahrgänge - zurechtfinden
mußte und daß Fortschritte in der Wissenschaft auch unter weniger
günstigen Umständen als den heutigen möglich sind.

Hans Hermann Weber wurde am 17. Juni 1896 in Berlin geboren. Er be-
suchte das humanistische Gymnasium bis zum Kriegsausbruch. Verwundet
begann er 1916 das Medizinstudium. Nach Kriegsende legte er das Staats-
examen 1921 in Rostock ab. Sein Doktorvater war Hans Winterstein, der
ihn in Fortsetzung eigener Arbeiten aus dem Jahre 1916 über die Rolle
der Milchsäure bei der Bildung und Lösung der Muskelstarre arbeiten
ließ. Weber hat Winterstein sehr verehrt. 1922 ging Weber für ein
halbes Jahr nach Kiel zu Otto Meyerhof, der dort als Assistent von
Rudolf Höber arbeitete. Weber erzählte oft, wie Meyerhof ihn als An-
fänger schließlich unter großem Vorbehalt akzeptierte, natürlich nur
als unbezahlten Assistenten. Bei Meyerhof hat Weber über die Oxida-
tionsvorgänge am Kohlemodell gearbeitet, einer Anregung von Warburg
folgend. Obgleich sich diese Arbeit als ein nur wenig zukunftsträch-
tiges Thema erwies, hat ihn dieser Aufenthalt in Meyerhofs Laboratorium
sehr beeindruckt und geprägt. Er kehrte zu Winterstein nach Rostock
zurück und begann sich mit den Grundlagen der Meyerhofschen Entioni-
sierungstheorie der Muskelkontraktion auseinanderzusetzen. Nach der
Habilitation 1925 ging Weber nach Berlin ans Pathologische Institut
zu Peter Rona. In der anregenden Atmosphäre dieser Zeit haben ihn
Fritz Haber, Leonor Michaelis und Otto Warburg nachhaltig beeinflußt.
Im Dahlemer Kolloquium hat er sich im wissenschaftlichen Streitge-

gespräch geübt, das er meisterhaft beherrschte und in dem ihm nur
schwer zu widerstehen war. 1927 fand er schließlich im Institut für
Physiologie und physiologische Chemie in Münster bei R. Rosemann eine
Assistentenstelle. Von 1933 an verwaltete er den neugeschaffenen Lehr-
stuhl für physiologische Chemie. Seine endgültige Ernennung durch das
Kultusministerium Berlin blieb jedoch aus, weil er in Münster als po-
litisch nicht zuverlässig galt. 1939 folgte er schließlich einem Ruf
nach Königsberg auf den Lehrstuhl für Physiologie und physiologische
Chemie. Kurz bevor Königsberg eingeschlossen wurde, konnte Weber mit
Hilfe eines Forschungsauftrages zur Gewinnung von Blutkonserven die
Stadt verlassen. In Tübingen fand er eine neue Wirkungsstätte. Hier
herrschte eine für die damalige Zeit exzeptionelle wissenschaftliche
Atmosphäre. Mehrere Berliner Kaiser Wilhelm-Institute hatten sich
nach Tübingen oder in seine Umgebung geflüchtet. Ein großes wissenschaftliches Ereignis in den ersten Nachkriegsjahren waren Besuche
amerikanischer Wissenschaftsdelegationen unter der Leitung von Otto
Krayer 1948 und Erwin Straus 1952, die von der unitarischen Kirche
Amerikas finanziert wurden. Die amerikanischen Unitarier bemühten
sich, den unterbrochenen wissenschaftlichen Kontakt zwischen Amerika
und Europa wieder herzustellen, indem sie renommierte amerikanische
Wissenschaftler zu Vorträgen nach Europa schickten. Weber koordinierte
diese Aktion in Tübingen, und in Anerkennung seiner neueren Arbeit
überreichte ihm E. Straus einen Scheck über 10.000 Dollar. Bei einem
Jahresetat des Instituts von 10.000 DM konnten damit die drückendsten
Mängel behoben werden. Ende 1953 erhielt Weber dann einen Ruf an das
Institut für Physiologie im Max-Planck-Institut für medizinische Forschung in Heidelberg. Sein Vorgänger, Hermann Rein, der das Institut
1952 übernommen hatte, war einer tückischen Krankheit erlegen. Mit der
Annahme des Rufes nach Heidelberg hat sich in Webers Leben ein Kreis
geschlossen. Er übernahm das Institut, dessen erster Direktor sein
verehrter Lehrer Otto Meyerhof gewesen war. Er war sich dieser Tradition immer bewußt, Ausdruck seiner Bewunderung und Verehrung für Meyerhof war seine Rede zur Eröffnung des Meyerhof-Symposiums 1970 in Heidelberg. In Heidelberg hat sich Webers wissenschaftliches Werk voll-
endet, an dem er über 40 Jahre konsequent gearbeitet hat.

Drei Perioden kennzeichnen seine wissenschaftliche Produktivität. Der
Periode des Myogens folgten die Studien über das Myosin, und diese
wurden abgelöst durch die Analyse der Wechselwirkungen des ATP mit dem
Aktomyosin. Nach der kurzen Zusammenarbeit mit Meyerhof 1922 konzen-
trierte Weber sich ganz auf das Studium des wasserlöslichen Proteins
des Muskels, des Myogens, das man damals für ein Protein sui generis
hielt. Er benutzte das Myogen als Modellsubstanz, an welcher er die
Voraussetzungen, die ein Muskeleiweiß, in dem chemische Energie in me-
chanische Spannung umgesetzt wird, erfüllen muß, analysieren wollte.
Da der Vorgang wegen seiner großen Geschwindigkeit einer physikochemischen Untersuchung nicht zugänglich ist, ist man auf Kombinationen
angewiesen, deren Grundlagen eine systematische Kenntnis nicht nur
der energetischen Vorgänge, sondern auch der physikalischen und che-
mischen Verhältnisse der Muskelgrundsubstanz ist. Sein Ziel war es,
die physikochemischen Grundlagen für die Meyerhofsche Muskelenergetik
zu schaffen. Die Analyse beginnt in vielen seiner Arbeiten mit der
Gegenüberstellung konträrer Hypothesen, wie hier der Fürthschen Quel-
lungstheorie und der Meyerhofschen Entquellungs- oder Entionisierungs-
hypothese. Die Entscheidung wird gesucht durch die osmometrische Be-
stimmung des isoelektrischen Punktes des Myogens. Er kommt zu dem
Schluß, daß die Säurequellung als Mechanismus der Kontraktion auszu-
schließen ist. Aus der Beobachtung Biedermanns, daß das Myogen im
Sarkoplasma lokalisiert ist, sah er keine Entwertung seiner Befunde.
Denn, so meinte er, die auf der Ionisierungskurve des Myogens aufge-
baute Ableitung dürfte prinzipiell für alle Eiweißkörper im Muskel
gelten, da alle Proteine polyvalente Ionen sind. Ein Höhepunkt in der

Myogenära war die Arbeit, in der er das Massenwirkungsgesetz auf die Proteine als Zwitterionen anwendet. Er und Lindström-Lang waren die Ersten, die nach den Anregungen von Bjerrum die Proteine als polyvalente Ionen behandelten. Oft hat er davon erzählt, wie er in den Ferien an der See die schwerfälligen Formeln der Arbeit abgeleitet hat.

In der ersten Arbeit über das Myosin, dessen Sonderexistenz neben Myogen Weber noch nicht mit Sicherheit erwiesen schien, untersuchte er sein elektrochemisches Verhalten und griff damit wieder in die Diskussion über die Meyerhofsche Kontraktionstheorie ein. Mit G. Böhm hat Weber 1931 an geordneten Myosinfäden die ersten Röntgendiagramme aufgenommen und 1941 mit M. v. Ardenne im Elektronenmikroskop die filamentösen Strukturen des Myosins beobachtet. Größte Anerkennung brachten ihm die Ergebnisse der Untersuchungen der polarisationsoptischen und mechanischen Eigenschaften seiner Myosinfäden. Sie führten zusammen mit einer Mengenanalyse der Muskeleiweißkörper zu dem Schluß, daß die Myosinstäbchen in anisotropen Abschnitten der quergestreiften Muskulatur lokalisiert sind. Diese Untersuchungen lieferten das Material für seinen klassischen Beitrag zu den Ergebnissen der Physiologie 1934. Er hat damals bereits vergeblich nach Wechselwirkungen der Zwischenstoffe des Muskelstoffwechsels mit seinen Myosinfäden gesucht. Als in Deutschland wissenschaftliches Arbeiten nahezu unmöglich geworden war, haben Albert Szent-Györgyi und seine Mitarbeiter die Weberschen Resultate aufgegriffen und durch unbefangenes Vorgehen die aufsehenerregende Entdeckung gemacht, daß es neben dem Myosin ein fädiges Protein im Muskel gibt, das mit Myosin Komplexe bildet, das Aktin. Darüber hinaus waren kurz zuvor in England durch Needham und in Rußland durch Engelhardt und Ljubimova Wechselwirkungen zwischen ATP und Myosin entdeckt worden, die die weitere Entwicklung maßgeblich beeinflußt haben. In diese stürmische Entwicklung hat Weber ohne Zögern eingegriffen, nachdem 1948 in Tübingen bescheidene Arbeitsmöglichkeiten gefunden waren. Zunächst glaubte er, daß es vordringlicher sei, die Molekulardaten des Myosins zu sichern. Diese Untersuchungen hat er mit Gerhard Schramm und Hildegard Portzehl durchgeführt. Dann hat er seine Aufmerksamkeit und die seiner Mitarbeiter auf die Wechselwirkungen des ATP mit den kontraktilen Proteinen, seine dissoziierende und seine synäretische Wirkung gelenkt. Er hat in Anlehnung an seine Erfahrungen mit den Myosinfäden die Herstellung hochgeordneter kontraktiler Aktomyosinfäden angeregt und die von A. Szent-Györgyi eingeführte Glyzerinextraktion des Muskels dazu benutzt, das kontraktile Protein in seiner natürlichen Anordnung einer kritischen Analyse zugänglich zu machen. Die anfänglichen Versuche, die Wirkung des ATP als eine reversible Verminderung der Kohäsionskräfte zwischen den Filamenten und die Kontraktion als ein kinetisch-entropisches Phänomen zu erklären, hat er schnell aufgegeben, nachdem die Beziehungen zwischen der Spaltung des ATP durch die kontraktilen Proteine und ihre mechanischen Veränderungen immer deutlicher wurden. Er hat dann das Studium zwischen ATP-Spaltung und mechanischer Leistung verschiedener kontraktiler Systeme vorangetrieben. Die Ergebnisse hat er so zusammengefaßt:

1. Im Ruhezustand ist die kontraktile Substanz sehr dehnbar und beinahe plastisch, weil der ruhende Muskel den Weichmacher ATP enthält, ohne ihn spalten zu müssen.

2. Bei der Arbeit kontrahiert sich das Aktomyosin, weil als Folge der Erregung ATP gespalten wird und das Aktomyosin erschlafft wieder, weil diese Spaltung aufhört, ehe der Weichmacher ATP in seinem Bestand erschöpft ist.

3. Falls es im intakten Muskel doch zu einer Erschöpfung des ATP-Bestandes kommt, wird der Muskel starr (Totenstarre).

In Heidelberg wurden dann die kontraktilen Proteine einfacher motiler
Strukturen, wie Fibroplasten - durch Hartmut Hoffmann-Berling - mit
in die Untersuchung einbezogen. Es wurden Arbeiten initiiert, die hel-
fen sollten, den Mechanismus der Energietransformation im kontraktilen
System zu verstehen. In Filmen aus dieser Zeit, die zeigen, daß sich
isolierte Myofibrillen unter der Wirkung des ATP verkürzen, sieht man,
daß die Kontraktion ohne Verkürzung der A-Banden verläuft. Wenn dennoch
die Sliding-filament-Theorie von A.F. Huxley, H.E. Huxley und J. Hanson
überraschte, so wohl deshalb, weil Weber 1941 an eine Verschiebung
längenkonstanter Elemente gedacht hat, sie aber damals nicht als Grund-
lage für die Verkürzung der Spannungsentwicklung des Muskels akzeptie-
ren konnte.

Webers Verdienst ist es, durch seine frühen Arbeiten für eine Konti-
nuität in der Erforschung der Muskelproteine gesorgt zu haben und damit
den Boden für die großen Fortschritte auf dem Gebiet der Physiologie
in den 50er und 60er Jahren bereitet zu haben. Seine bedeutendste
wissenschaftliche Leistung war die Aufdeckung der Beziehungen zwischen
der Hydrolyse des ATP durch die kontraktilen Proteine und ihrer mechanischen Leistung. Als Dank haben ihm Schüler und Freunde aus aller
Welt zum 70. Geburtstag einen Band im vorletzten Jahrgang der Biochemischen Zeitschrift gewidmet.

Webers Stellung zur deutschen Physiologie und physiologischen Chemie
war gekennzeichnet durch eine gewisse Ambivalenz. Im Gegensatz zu
angelsächsischen Wissenschaftlern zeigten deutsche Physiologen und
physiologische Chemiker lange Zeit kein besonderes Verständnis für
sein wissenschaftliches Vorgehen. Das hohe internationale Ansehen,
das er nach dem zweiten Weltkrieg in der Welt fand, hat ihm schließlich auch in Deutschland Anerkennung gebracht. Mit diesem Symposium
würdigt die Gesellschaft für biologische Chemie Hans Hermann Weber
und sein wissenschaftliches Werk.

W. Hasselbach

Mechanism of Chemo-mechanical Energy Transformation

The Structural Basis of Contraction and Regulation in Skeletal Muscle

H. E. Huxley

A. Introduction

The purpose of this review will be to present sufficient of the background information about the structure of muscle, as well as some details of recent work, so that a non-specialist can form a reasonably coherent picture of the present state of our knowledge, of the problems which have to be solved, and of some of the current lines of experimentation.

There are three reasons why this field is a particularly interesting one at the present time. First, the knowledge that we now have, though it is derived from a number of rather different experimental approaches - physiology, biochemistry, enzyme kinetics, structural analysis - is sufficiently comprehensive that different areas of the jigsaw puzzle are beginning to come together - to fit together in some places, and to reveal defects and incompatibilities in others. Secondly, for a variety of obvious reasons, muscle provides a favourable situation for studying many problems of wide biological interest - problems of enzyme action, of the assembly of macromolecular structures, of the transduction of energy from one form to another, even of the structure and function of membranes. Thirdly, it has emerged in the last few years that proteins almost identical to the major proteins in muscle are present in a wide variety of other cell types, where they are involved in a number of different kinds of cell movement. The structures involved in producing these movements are much less highly organised than those in muscle, but there are very strong grounds for believing that the underlying processes involved are the same (Huxley, 1973). Previous experience with muscle proteins and their interactions which has been facilitated by the ready availability of abundant and highly structured material, can be of great help in understanding these considerably less accessible and abundant cellular motile systems.

B. Basic Structural Features of Muscle

My first task is to describe the fundamental structural information we have about the contractile system in striated muscle. To a large extent, I will have to do this without discussing the evidence for the various features that I will point out, simply because it would take too long. There is now a large measure of agreement about most of these features, and I will attempt to be as objective as I can over questions where there are still significant inadequacies in the present evidence.

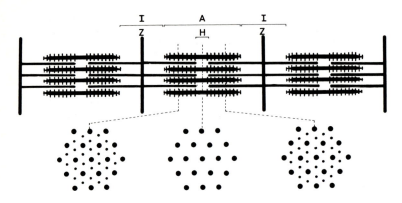

Fig. 1. Diagrammatic representation of construction of striated muscle
from overlapping arrays of thick (myosin) and thin (actin-tropomyosin-
tropin) filaments. Sliding force between filaments is generated by
repetitive cyclic movement of cross-bridges. Attachment of cross-
bridges to thin filaments is blocked by regulatory system when muscle
is switched off

All the available evidence is strongly in favour of a sliding filament mechanism in which the proteins actin and myosin are organised
in the myofibrils into separate but partially overlapping arrays of
filaments (Fig. 1) which slide past each other and increase the extent
of overlap during shortening of the muscle. This basic model was originally proposed in 1954, independently, by A.F. Huxley and R. Niedergerke and by myself and the late Jean Hanson. The sliding force between the actin and myosin filaments is believed to be generated by
cross-bridges projecting outwards from the myosin filaments, attaching in a cyclical fashion to actin, as suggested by Hanson and Huxley
in 1955, splitting ATP as they do so and thereby releasing the energy
for contraction. These cross-bridges represent the enzymatically active parts of the myosin molecule.

Myosin is a molecule with a very remarkable structure (Lowey et al.,
1969). Basically it contains two very large polypeptide chains of
molecular weight about 200,000 daltons each (the "heavy chains") and
four smaller polypeptide chains having molecular weights in the 20,000
daltons range. Along part of their length, the two heavy chains are
coiled around each other to form a 2-chain α-helical coiled-coil structure about 1,400 Å in length and 20 Å in diameter. About one half of
each heavy chain is involved in this structure. The rest of each heavy
chain is folded up separately in a globular form, together with some
or all of the light chains. The two heavy chains are of very similar
amino-acid sequence, and are arranged in parallel to each other with
the same polarity so that the two globular regions are located at the
same end of the molecule. The α-helical portions of the myosin molecules are involved in forming the backbone of the thick filaments, by
side-to-side bonding along part of their length with their neighbours,
whereas the globular regions - known as the S_1-subunits - which have
on them the sites for splitting ATP and combining with actin - project
out sideways from the thick filaments and form the cross-bridges. Present evidence indicates that a portion of the rod part of myosin
(known as the S_2 region) can hinge out sideways from the backbone of
the thick filaments so as to allow the 'head' or S_1 subunits to attach
to the actin filaments alongside, whose sidespacing from the myosin
filament backbone varies somewhat according to muscle length. The

thick filaments are about 1.6 microns in length and lie about 400 Å apart. Each of them contains about 250-300 myosin molecules, which corresponds to a concentration of approximately 10^{-4} M.

The thin filaments contain actin, a protein of molecular weight about 42,000 which forms globular units approximately 50 Å in diameter which in turn assemble into filaments composed of two helically wound strings of the G-actin units. The thin filaments in vertebrate striated muscle also contain the regulatory proteins troponin and tropomyosin, which are involved in switching the actin-myosin interaction on or off in response to changes in calcium-concentration.

C. Behaviour of Cross-bridges

In a resting muscle, the cross-bridges are not attached to actin, the filaments can slide past each other readily under an external force, and the muscle is plastic and readily extensible. When a muscle contracts, it is supposed that any particular cross-bridge will first attach to actin in one configuration, in an approximately perpendicular orientation and then, while still attached, will undergo some configurational change so that its effective angle of attachment alters, i.e. it 'swings' or 'tilts' in such a direction as to pull the actin filament along to the direction of the centre of the A-band (Fig. 2) (Huxley, 1969). When this movement is complete - the extent of movement probably being 50-100Å - the cross-bridge can be detached from actin by the binding to it of another molecule of ATP (whose first effect is to dissociate actin and myosin). The ATP is then split by the myosin head, while still uncombined with actin (as indicated by the work of Lymn and Taylor, 1970) but the reaction products remain attached to the enzyme, with the complex probably in a "strained" state, until the enzyme once more attaches to actin and releases the stored energy in the form of mechanical work. The combined effect of all the cross-bridges undergoing these asynchronous cycles of attachment, pulling and detachment is to produce a steady sliding force which will continue as long as the muscle is active. It will be realised that this type of mechanism depends on a very specific interaction between actin and myosin molecules and therefore requires that they be built into the structure with very specific orientations. It is found in practice, from electron microscope observations, (Huxley, 1963) that the myosin molecules along one half of the length of each thick filament (and hence in one half A-band) are all oriented with their 'tails' pointing in one direction (towards the centre of the A-band). In the other half of the filament, this polarity is reversed, so that the tails again point towards the centre. Such an arrangement would ensure that all the elements of force generated by individual cross-bridges will add up in the proper direction. Similarly, all the actin monomers along each thin filament have the same structural polarity (which reverses at the Z-lines) so that they will all be able to interact with myosin cross-bridges in identical fashion. Indeed, one might regard one of the most essential features of the structure of muscle as being the organisation of all the individual interacting molecules in a large body of tissue so that they act in a concerted fashion, both spatially and temporally.

This would be an appropriate point at which to mention the mechanism by which all the contractile material in a whole muscle can be switched on or off in a small fraction of a second - even in a few milliseconds in some cases.

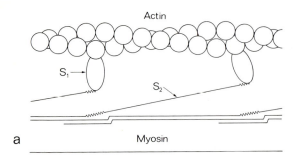

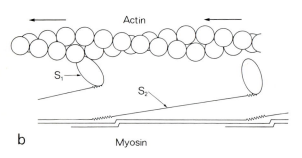

Fig. 2a and b. Active change in angle of attachment of cross-bridges (S_1 subunits) to actin filaments could produce relative sliding move-ment between filaments maintained at constant lateral separation (for small changes in muscle length) by long range force balance. Bridges can act asynchonously since subunit and helical periodicities differ in the actin and myosin filaments. (a) Left-hand bridge has just at-tached; other bridge is already partly tilted. (b) Left-hand bridge has just come to end of its working stroke; other bridge has already detached, and will probably not be able to attach to this actin fila-ment again until further sliding brings helically arranged sites on actin into favourable orientation

Following the arrival of nerve impulses down the motor nerve, trans-mission of chemical signals takes place across the motor end plate and these in turn bring about a depolarisation of the external membrane of the muscle. This depolarisation is, in skeletal muscle, transmitted inwards as an electrical signal along invaginations of the external membrane which form the so-called transverse tubules, and which extend throughout the interior of the fibre. It is still not completely clear whether the depolarisation is normally transmitted by passive spread, or as a regenerative action potential, but in either case the trans-mission time will be only of the order of a millisecond. The trans-verse tubules form specialised junctions with another component of the sarcoplasmic reticulum, the longitudinal system, which has the form of flattened membrane-bound sacs lying in between the myofibrils and some-times forming an almost complete sheath around them. In amphibian muscle (e.g. frog sartorius) the junctions occur at the level of the Z-lines, one transverse tubule lying between the longitudinal elements associated with two adjoining sarcomeres, and the junction is known as a triad. In vertebrate muscle, the junctions are located near the A-I boundaries. The reticulum shows up well in the electron-microscope and has been de-scribed by a number of authors (e.g. Porter and Palade, 1957; Andersson-Cedergren, 1959; Franzini-Armstrong and Porter, 1964). Its function has also been well reviewed (Ebashi and Endo, 1968; A.F. Huxley, 1971).

Calcium is sequestered in the longitudinal elements of the reticulum, and in a resting muscle the concentration of free calcium in contact with the myofibrils is kept down to a value in the region of 10^{-7} M by the very efficient calcium pump associated with the reticulum. Following stimulation, calcium is rapidly released from the reticulum so that the concentration of free calcium ions in contact with the myofibrils rises to the region of 10^{-5} M. Since the calcium has to diffuse a distance of only a micron or less, the rise in calcium concentration could follow the release with a delay of only about a millisecond. How the signal for the release of calcium is transmitted from the transverse tubules to the longitudinal reticulum (where the calcium is stored) is not known at present, and is a difficult problem to investigate directly because of the small size of the structures involved. This calcium is the chemical signal for contraction and it switches on the contractile activity of the myofibrils when it reaches them. Thus, in principle the signal for contraction can be transmitted from the outside membrane to the whole of the contractile material within a few milliseconds. This is obviously advantageous to an animal which may want to start to move rather quickly! Mechanisms dependent on the diffusion inwards of some substance from the outside membrane would be much too slow to account for the known rates at which muscles can become fully activated, as A.V. Hill pointed out many years ago (Hill, 1948, 1949). I will consider later the details of the mechanism by which calcium activates the contractile mechanism. First of all though, we should consider the nature of the cross-bridge action it-self in greater detail, and see what are the present outstanding problems.

The information we have about the cross-bridges is limited and fragmentary. The existence of some form of lateral cross-connection between separated filaments of myosin and of actin was originally proposed (Huxley, 1952) in order to account for the greatly increased resistance to stretch exhibited by a muscle in rigor, and this argument still remains a very powerful one. The visualisation of cross-bridges in electron micrographs of muscle was first described in 1953, and their appearance was shown rather more clearly a few years later (Huxley, 1953, 1957; Huxley and Hanson, 1956); the actual images of cross-bridges in muscle have not improved much since that time. They appear as projections originating on the thick filaments, having a diameter of about 50 $\overset{\circ}{A}$ and a length of about 100-150 $\overset{\circ}{A}$, extending out towards the thin filaments and attached to them in muscles in rigor. In insect indirect flight muscle in rigor they can be seen to be attached to the actin filaments in a characteristically tilted configuation (Reedy et al., 1965) the tilt being in such a direction as to move the attached end of the cross-bridge towards the centre of the A-band. The attachment of 'free' myosin heads can be examined in the electron microscope by using the negative staining technique to examine actin filaments 'decorated' with myosin subfragment 1 (S_1) in absence of nucleotide (Huxley, 1963; Moore et al., 1970) and a tilted form of attachment is again apparent. It is very likely, therefore, that this corresponds to the configuration adopted by the cross-bridge at the end of its working stroke, when ADP and P_i have been released, and when no further force is being exerted. The X-ray and EM evidence indicate that bridges in relaxed muscle are approximately perpendicular to the filament axis, so the simplest supposition would be that they attached to actin in this configuration at the beginning of their working stroke. The extent of movement per stroke that this would imply (based simply on geometric considerations) is about 70-80 $\overset{\circ}{A}$, which accords well with the values obtained by Huxley and Simmons (1971, 1972) from observations on rapid mechanical transients in muscle. The return stroke of the cycle does not necessarily require much energy to be expended - it could simply correspond

to return to the equilibrium configuration adopted by a cross-bridge
when carrying ATP or its immediate split products. It should be recalled that even in the absence of actin (in a very stretched muscle)
the cross-bridges lose their regular arrangement in the absence of
ATP and it seems likely, but has not been established, that under
these conditions they no longer adopt a constant and approximately
perpendicular angle of tilt.

One of the best, though by no means straightforward methods of in-
vestigating the nature and behaviour of the cross-bridges is by low-
angle X-ray diffraction, since a large part of the diagram comes from
the cross-bridges themselves, and since it is possible to study a
muscle by this technique under almost normal working conditions. The
disadvantage of this technique - besides the inherent and well-known
ambiguities of X-ray diagrams! - is that the reflections from muscle
are rather weak (about one millionth as strong as the direct beam) -
and it has therefore been necessary to invest a considerable amount
of time and effort into the technical innovations required to record
the patterns sufficiently rapidly. This is not an appropriate occasion
to discuss these developments; suffice it to say that we can now record changes in the strongest reflections with a time resolution of
10 milliseconds; but we could still make use of gains of X-ray intensity by several orders of magnitude in order to measure weaker reflections and to avoid having to use very long series of contractions.
However, what of the results? In the diagram from a resting muscle
there is a well developed system of layer lines with a 429 $\overset{o}{A}$ axial
repeat and a strong third order meridional repeat at 143 $\overset{o}{A}$ (Huxley
and Brown, 1967; Elliott et al., 1967). This pattern arises from a
regular helical arrangement of cross-bridges on the thick filaments,
with groups of cross-bridges accurring at intervals of 143 $\overset{o}{A}$ along
the length of the filaments with a helical repeat of 429 $\overset{o}{A}$. The number
of cross-bridges in each group is not yet absolutely certain, but it
is more likely to be three than two (Squire, 1973).

Diagrams from contracting muscle may be recorded on film using a shut-
ter so as to transmit the X-ray beam only when the muscle is being
stimulated. A long series of tetani is necessary, with intervals in
between them for recovery - usually one second tetani and two minute
intervals. Such diagrams show that the whole pattern becomes very
much weaker during contraction, indicating that the cross-bridges are
much less regularly arranged, as would be expected if they were undergoing asynchronous longitudinal or tilting movements (and possibly
lateral ones too) during their tension-generating cycles of attachment
to actin. If this is indeed the case, then the very rapid development
of the active state of a muscle following stimulation should be accompanied by an equally rapid decrease in intensity of the layer line
pattern. Recent technical developments have now made it possible to
record the X-ray diagram sufficiently rapidly to investigate this
question. Studying single twitches of frog sartorius muscle, Dr. John
Haselgrove and I have found (1973) that the change in pattern (at 10°C)
begins about 10-15 msec after stimulation and is half-complete by
about 20 msec, a time when the externally-measured tension (which in
a normal isometric contraction is delayed behind the onset of the
active state by the necessity to stretch the series elastic elements
before the internal activity can manifest itself as tension) has hardly begun to rise.

The equatorial part of the X-ray diagrams from muscles is also very
informative. It is generated by the regular side-by-side hexagonal
lattice in which the filaments are arranged, and the relative intensity of the two principle reflections is strongly influenced by the
lateral position of the cross-bridges. Large changes occur as between

resting muscle and muscle in rigor (Huxley, 1952, 1953a, b, 1968) when
a high proportion, if not all, of the cross-bridges will be attached
to actin. These have been interpreted as indicating that in a resting
muscle, the cross-bridges lie relatively closer to the backbone of
the thick filaments, whereas when they attach to actin they hinge
further out and lie with their centres of mass nearer to the axes of
the actin filaments at the trigonal positions of the hexagonal lattice.

Similar changes have been observed in contracting muscles (Haselgrove
and Huxley, 1973), though the extent of change is less and would correspond to about half the cross-bridges being in the vicinity of the
actin filaments at any one time. Again, the observations are consistent
with a model in which the cross-bridges are undergoing a mechanical
cycle of attachment to and detachment from the actin filaments during
contraction. However, it should be appreciated that the parameter that
is being measured is the average lateral position of the cross-bridges
and this provides no direct evidence concerning the proportion attached,
or even indeed whether any are attached at all. It does not follow -
indeed it is very unlikely - that the average angle of attachment of
the cross-bridges to actin is the same during contraction as it is in
rigor, and this angle will affect the position of the centre of mass
of the cross-bridge relative to the axis of the actin filament. So
the proportion attached may be more than 50%; or it may be less if
some cross-bridges lie near to actin but are not themselves attached.

Nevertheless, the changes do indicate that a substantial lateral movement of the cross-bridges takes place during contraction and, as in
the case of the layer-line changes, it is important to establish
whether the movement occurs at a sufficiently rapid rate for it to
arise from a force-generating attachment of cross-bridges. Again,
this is a quantity that we can now measure, and Dr. Haselgrove and I
have found that the expected changes in the equatorial X-ray diagram
(decrease in intensity of [10] reflection, increase in intensity of
[11] reflection) do indeed occur with great rapidity after stimulation.
At 10°C, for example, the change is half complete within about 20 msec,
according well with the expected temporal characteristics of the active state.

There are other notable features of the X-ray diagram which give information - or at least clues - about cross-bridge behaviour. When
a muscle goes into rigor, for example, the axial X-ray diagram loses
all the features associated with the myosin filament helix and shows
instead a pattern of reflections which can be indexed on the actin
helix. This shows that the cross-bridges, the S_1 heads of myosin, are
relatively flexibly attached to the myosin filament backbone, but
from a relatively rigid attachment (in rigor anymay) to actin. The
same conclusion can be drawn from a comparison of the very regular
and highly ordered appearance in the electron microscope of negatively
stained specimens of actin filaments decorated with S_1 and the very
disordered appearance of the cross-bridges on myosin filaments examined by the same technique (Huxley, 1963, 1969). This behaviour not
only makes it easy now to understand how an absolutely constant structural cycle for the myosin head-actin interaction (presumably a requirement for these two proteins to interact enzymatically) can be
combined with a variable side-spacing between the actin and myosin
filaments. More important still, the characteristics of the two types
of attachment suggest very strongly either that the site at which
force originates is at the interacting surface of the myosin head and
the monomer on the actin filament, or that a rigid attachment is formed
there and is followed by a change in shape of the myosin head which
alters its effective angle of attachment. Forceful cross-bridge move-

ment generated by linkages between the S_1 subunit and the backbone of the myosin filament seems much less plausible.

Another feature of the diagram is that in a stretched muscle - even in one stretched so that there is no overlap between the arrays of actin and myosin filaments - the characteristic resting layer-line pattern of the myosin filaments is lost when the muscle goes into rigor, even in the absence of calcium. Thus, the presence of ATP (or its split products) on the myosin heads is a requirement for them to form a regular arrangement around the myosin filament backbone, though they will maintain such an arrangement in the presence of ATP and the absence of actin. The resting arrangement does not seem to be re-stored, even in stretched muscle, by unhydrolysable analogues of ATP such as AMP.PNP (Lymn and Huxley, 1975).

A further unexpected observation on muscles stretched beyond no-over-lap is that the decrease in intensity of the axial reflections during contraction still seems to occur. This finding should be interpreted with some caution (Huxley, 1972), but obviously one interpretation is that there exists a calcium-sensitive switch on the myosin filaments, which allows "actin-searching" cross-bridge movement to occur, (as well as the calcium-sensitive switch on the actin filaments, which allows actin-myosin combination to occur) and that this 'searching' movement takes place even when actin is not accessible.

In general, then, while we are fortunate that the cross-bridges do show up so well in the different features of the X-ray diagram, many things about them still remain unclear. Most notably, we have virtual-ly no information about the configuration in which they attach to actin when the myosin is still loaded with ATP or its immediate split products, at the beginning of the working stroke, and we have virtual-ly no information about the precise form of that working stroke. Fur-thermore, we have no detailed structural information about the myosin head subunits themselves, since they have still not been crystallised and so are not available for crystallographic analysis.

D. Structural Evidence about the Regulatory Mechanism in the Thin Filaments

In the light of the evidence about cross-bridge behaviour, incomplete as it is, we believe that we are nevertheless on fairly secure ground in supposing that the development of the active state in a muscle is characterised by the rapid attachment to actin of the myosin cross-bridges. It is apparent, therefore, that the calcium switch must con-trol whether or not this attachment takes place. In low concentra-tions of calcium it does not, in higher concentrations (around 10^{-6}- 10^{-5} M) it does. Moreover, one can see very readily that there are strong reasons why any control mechanism would have to interrupt the cycle at this particular point. While a muscle's first function is to produce movement by applying a force, it also has to allow that move-ment to be reversed, either by other muscles or by an external force at some subsequent time. Thus when a muscle is not actively contracting, it must be kept in a state where it offers a minimum of resistance to passive length changes. If the cross-bridges between the actin and myosin filaments were attached, but not cycling, the muscle would be in a rigid, inextensible state - indeed this is believed to correspond to the condition of rigor. Thus it is vital that the switch operates at such a point in the cycle that when the muscle is switched off, bridges are allowed to detach but are prevented from reattaching. Furthermore, if attachment to actin for one part of the cycle is required for the continued splitting of molecules of ATP to take place, then the energy producing reactions will automatically be switched off at the same time that the mechanical manifestations of contractile activity cease.

How, then, does calcium effect this transition between rest and activity?

In this brief summary, I will not attempt to give a full account of the development of the various lines of evidence that have led to our present views (see, for example, excellent reviews by Ebashi and Endo, 1968; Weber and Murray, 1973). Instead, I will summarise the basic features of what seems to me the simplest plausible model with particular reference to the structural evidence.

Control by calcium of the contractile activity of the actin-myosin system in vertebrate striated muscle requires the presence of the regulatory proteins troponin and tropomyosin. These proteins form part of the structure of the thin filaments (Hanson and Lowy, 1963; Pepe, 1966; Ohtsuki et al., 1967). Troponin is a protein complex consisting of three different subunits (one copy of each); one of the subunits binds calcium very tightly, but all three together are required for regulation. Troponin is located along the length of the thin filaments at intervals of approximately 385 Å and the stoichiometry is such that there is one troponin complex and one tropomyosin molecule for every seven actin monomers (Potter and Gergely, 1974). This stoichiometry is particularly significant if it is recalled that the actin filaments consist of two chains of actin monomers twisted round each other with a helical repeat of 360-380 Å and with a subunit repeat along each chain of 54.6 Å, so that there are approximately seven actin subunits along each chain within each helical repeat, and therefore exactly seven actin subunits per chain for each (approximately) 385 Å troponin repeat. Troponin appears to be a relatively globular molecule, but tropomyosin is a two chain coiled-coil α-helical structure present in sufficient amounts to provide two continuous strands running along the length of the actin filaments and hence able to make contact with each actin monomer (Fig. 3). Since the presence of tropo-

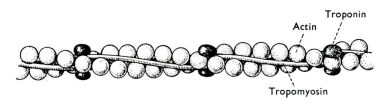

Fig. 3. Diagram showing probable general features of arrangement of actin, tropomyosin and troponin in the thin filaments of muscle (Hanson and Lowy, 1963; Ebashi and Endo, 1968)

myosin is necessary for the regulatory mechanism to operate, it is natural to suppose that the influence of troponin is transmitted to all the actin monomers via the tropomyosin strands. It is of interest, therefore, to see whether structural information about the thin filaments can provide any clues as to how this mechanism operates.

The first clue came from three-dimensional reconstruction studies of the mode of attachment of the myosin S_1 subunits to the actin monomers, using the so-called "decorated" actin (i.e. actin mixed with S_1 in absence of ATP) which probably corresponds to the "rigor" configuration of the cross-bridges and to their probable position at the end of their working stroke. Moore et al. (1970) found that the S_1 subunit, besides being tilted and skewed in a characteristic manner, was attached to the actin monomers in a somewhat tangential fashion, so that the end of the S_1 subunit and the contact area between it and the actin extended round into the groove in the actin structure. This can be seen in Fig. 4, showing an end-on projection of a short length of the structure.

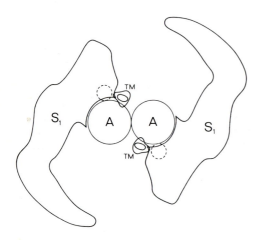

Fig. 4. Composite end-on view of actin-tropomyosin-S_1 structure. The shape of the S_1 subunits and the position at which they attach to the actin is copied directly from Plate IX of Moore et al. (1970). The actin structure (24 Å diameter spheres placed at a radius of 24 Å) is that found by Haselgrove (1972) to give the best agreement (for a simple model) with the stronger features of the X-ray diagram. The two tropomyosin (TM) positions correspond to: solid contours, activated state taken directly from difference Fourier (Plate X,b) of Spudich et al. (1972); dotted contours, relaxed state, based on radial position of reflection on third layer line of relaxed muscle. (This assumes that the "sense" of the azimuthal position of tropomyosin with respect to the attachment site of S_1 has been chosen correctly, which has not yet been proved.) The possible way in which tropomyosin could block the attachment of a cross-bridge is very evident

It must be recalled that the formation of "rigor links" (i.e. in absence of ATP) between myosin and actin is not inhibited by the regulatory complex even in the absence of calcium, and therefore attachment of the type seen in Fig. 4 would not be inhibited by the troponin-tropomyosin complex.

Nevertheless the projection of myosin into the groove is very suggestive and it seems at least a working possibility that when myosin attached to actin in the presence of ATP (or its split products), the end of the molecule projects a little bit further into the groove and that the attachment is vulnerable to the influence of tropomyosin. For the tropomyosin to just make contact with the myosin head in the rigor attachment, it would have to lie at an angle of about 60-70° to the respective actin monomer, measured from the helix axis. Fig. 4

shows the position of the rigor links, together with the outline of
the probable positions of tropomyosin given by the difference Fourier
of Spudich et al. (1972) (activated state) and from the X-ray data on
resting muscle (relaxed state), as discussed later. Spudich et al.
were able to reconstruct the complex filaments and show that a fairly
continuous strand of material ran along in each of the two grooves in
the actin helix, situated asymmetrically, so that if the actin structure is thought of as two strings of monomers twisted around each
other, one regulatory strand is more closely associated with one string
of monomers, and the other with the other string. The angular position
of each strand, measured from the axis of the helix, is about 60-70°
relative to the associated actin monomer at the same level. In these
reconstructions the strong 385 Å axial periodicity believed to arise
from the troponin component was not included, so that the resultant
structure would be expected to show the location of the tropomyosin
only. Thus the tropomyosin appears to run alongside the chains of actin
monomers in a position very suggestive of possible interaction. The
electron microscope evidence has been taken a stage further by the
work of Wakabayashi et al., 1975. These workers were concerned to see
whether a difference in the position of tropomyosin could be detected
in the electron microscope images of thin filaments in the active
and in the inhibited state. Since control of the calcium level could
not be assured during the negative staining process, and since much
more satisfactory paracrystalline arrays of filaments (needed for
good three-dimensional image reconstruction) were formed in the presence of calcium than in its absence, the inhibited state of the
filaments was produced by including only troponin-I (inhibitory subunit)
and troponin-T (tropomyosin combining subunit), together with tropomyosin, in the regulatory complex with which actin was combined. In
these circumstances, the filaments are in the 'switched-off' state
(i.e. unable to interact with myosin) whether calcium is present or
not, but it is reasonable to suppose that the mechanism of inhibition
is the same as that involved in the absence of calcium, when the calcium-binding subunit (Troponin-C) is present. The structure of the
inhibited filaments was compared with that of actin plus tropomyosin
in the absence of troponin, when the filaments are always in the
'switched-on' state. It was found that a very marked difference in
structure did indeed occur between the two different forms. In the
active filaments, the tropomyosin strands made relatively loose contact with the actin monomers, and though still asymmetrically disposed
in the long pitch grooves between the two 'strings' of actin monomers,
lay relatively closer to the centre of those grooves. In the inhibited
filaments, the tropomyosin strands made very close contact with the
actin monomers, in a position which could plausibly block the attachment of myosin, and further away from the centre of the long pitch
helical grooves. The results therefore provide further strong support
for the idea that regulation is associated with a significant lateral
movement of the tropomyosin strands, and may be effected by a relatively simple and straightforward mechanism by which tropomyosin physically blocks the attachment of myosin; the actin structure itself undergoes little or no change.

E. X-Ray Evidence

This model is strongly supported by X-ray diffraction evidence (Huxley
and Brown, 1967; Huxley, 1970, 1971a, b; Vibert et al., 1971, 1972).
The pitch and subunit repeat of the actin filaments remain virtually
constant in an actively contracting muscle, and the main pattern of
actin reflections do not alter their relative intensities either,

indicating that the major part of the actin filament structure remains unchanged too. As was pointed out at the time, this does not rule out repetitive cyclical changes taking place in a small part of the thin filament structure at any given time; but since calcium will be tightly bound to the troponin of the thin filaments for the whole period of activity, the changes it produced should be maintained virtually continuously throughout contraction and hence they cannot be of a kind which affects the helical parameters of the thin filaments. This view is strengthened by the observation that even in rigor (when the level of free calcium is normally high) the subunit repeat in the actin filaments is unchanged, and any change in the pitch of the actin helix is either relatively small or absent. Experience with other systems (e.g. haemoglobin) indicates that even relatively small internal changes in structure inside a molecule or subunit are liable to produce appreciable changes in the way the subunits pack together, and I therefore believe that the most likely interpretation of the constancy of the actin structure is a virtually complete constancy in the internal structure of the actin monomers thermselves when they are switched on or off. This strongly favours the possibility that regulation is affected by a steric blocking mechanism involving tropomyosin movement (Huxley, 1970, 1971a, b).

Positive evidence for tropomyosin movement comes from X-ray diffraction observations of some of the relatively weak parts of the diagram which arises from the thin filaments in muscle. Whilst, as I have already mentioned, the main parts of that diagram remain unaltered during contraction, there are reflections, rather far away from the meridian, on the second and third layer-lines (specifically at axial spacings of approximately 190 $\overset{\circ}{A}$ and 130 $\overset{\circ}{A}$ and radial spacings of about 0.021 $\overset{\circ}{A}{}^{-1}$ and 0.026 $\overset{\circ}{A}{}^{-1}$ respectively) in which very significant changes in intensity (but not layer line spacing) can be seen. There is a marked decrease in intensity of the third layer line reflection in a contracting muscle and a marked increase in intensity on the second layer line. There are strong technical grounds for believing that these reflections arise in very large part from the tropomyosin component of the thin filaments, and that the changes in them indicate a change in the position of the tropomyosin. These arguments can be made rather precise and rigorous (Haselgrove, 1972; Huxley, 1972; Parry and Squire, 1973), and it can be shown that the changes in the X-ray diagram upon contraction can be very well accounted for by a movement of tropomyosin, away from a position which might block myosin attachment to one lying 10-15 $\overset{\circ}{A}$ nearer the centre of the long pitch groove, a similar movement to that indicated by the 3-D reconstruction results. There are several attractive features of this type of mechanism and I will mention three of them very briefly. First of all, there is the basic one, that if the tropomyosin strand moves as a whole, and possibly as a relatively rigid structure (it is a two chain coiled-coil molecule) it is easy to see how one tropomyosin molecule could regulate the seven actin monomers over which it extends. Secondly, in such a model, what happens at one myosin-binding site on actin (say attachment of a myosin head not loaded with ATP) can influence what happens on adjoining ones, by causing a displacement of tropomyosin. This could account for some of the very remarkable cooperative effects in the interaction, as discussed by Weber and Bremel (1972). Thirdly, the somewhat non-specific nature of the interaction that would be required between tropomyosin and the successive actin monomers would accord very well with the known features of the amino-acid sequence of tropomyosin (Sodek et al., 1972) in which there is not a precise sequence of amino-acid residues which repeats at regular intervals along the coiled-coil structure, but merely a tendency for similar groupings of residue to recur with the actin period (MacLachlan et al., 1975). One might imagine that a control mechanism mediated by a structural change _within_ the actin

monomers and transmitted along by tropomyosin would call for a very
precise and exact molecular interaction between the two points at each
successive actin monomer repeat; and this is not found. Additionally,
the fact that hybrid systems, involving types of actin which are not
associated with troponin *in vivo* (e.g. molluscan actin) can be regulated
by vertebrate troponin-tropomyosin again suggests that an absolutely
exact and specific pattern of interaction between each actin monomer
and tropomyosin is not a necessary requirement for regulation.

F. Control of Tropomyosin Movement

Finally, we must discuss whether there are any clues as to how the
required movement of tropomyosin might be brought about. Actin and
tropomyosin combine together quite strongly even in the absence of
troponin, but in the complex so formed, actin retains its ability to
activate myosin ATPase - indeed in some circumstances (Weber and
Murray, 1973) this ability may be enhanced. Tropomyosin, is in these
circumstances, held along the actin helix in such a way that a high
proportion, if not all, of the myosin-binding sites on actin are still
available. The intact troponin complex will combine not only with
actin-tropomyosin together but with actin and with tropomyosin separately. Thus if the same interactions are involved, actin and tropomyosin must, in the region where troponin attaches, be held in a
specific configuration relative to each other by bonds additional to
those between actin and tropomyosin directly. Accordingly, one can
reasonably envisage a model in which, in the absence of calcium, tropomyosin is held in the blocking position, and where some structural
change within the troponin complex occurs in the presence of calcium
which allows or causes tropomyosin to move towards the centre of the
long pitch grooves in the actin double helix and away from the position which sterically blocks myosin attachment (Huxley, 1972).

Very strong positive evidence in favour of such a model has been found
by Hitchcock et al. (1973), who studied the binding to actin-tropomyosin of a combination of troponin C plus troponin I, i.e. in the
absence of the third part of the troponin complex, troponin T, which
binds strongly to tropomyosin. In the absence of calcium, troponin
[I+C] binds strongly to actin-tropomyosin filaments (though it does
not, when present in amounts equivalent to the normal stoichiometry,
inhibit their ability to activate myosin ATPase). In the presence of
calcium, however, at concentrations [10^{-5} M] just sufficient to switch
off inhibition in a fully reconstituted system, troponin [I+C] de-
taches from actin-tropomyosin filaments and can be separated from them
by ultracentrifugation. This finding suggests a so-called 'two-site
model' (Hitchcock et al., 1973) in which troponin complex binds to
actin-tropomyosin via two separate sites on troponin. One of these,
on the T subunit, binds permanently (i.e. in both the presence and
absence of calcium) to tropomyosin, so that the intact troponin complex always remains attached to the thin filaments. The other binding
site is calcium-sensitive and attaches to actin-tropomyosin in the
absence of calcium only. The formation of this link holds tropomyosin
in the blocking position, providing the troponin T subunit is present.
When calcium is present, the link is broken, and tropomyosin can move
to a non-blocking position. Again, we have arrived at a rather simple
steric model.

Margossian and Cohen (1973) also support the same type of model, and
have observed an increase in the strength of the binding between
troponin C and troponin T in the presence of calcium which could re-

flect a change in the state of the troponin complex associated with tropomyosin movement.

Thus, in general, there are several interesting pieces of evidence which fit plausibly into a model for tropomyosin movement mediated by troponin. However, the real details of such a model have still to be worked out, and the model can only be accepted as a useful working hypothesis until that is done.

I should also mention that not all muscles are regulated by the troponin-tropomyosin system, though as far as we know they are all regulated by changes in the concentration of free calcium ions, over similar ranges of values. Thus molluscan muscle, for example, (Kendrick-Jones et al., 1970) is regulated by calcium sites on the myosin component, and lacks troponin, and certain other species even have both actin-linked and myosin-linked regulatory systems present in muscles at the same time (Lehman et al., 1972), though this has not so far been shown to be the case for vertebrate striated muscle.

In summary, then, there is a wealth of structural evidence about the mechanism of muscular contraction and its control. Much of this has been very useful in helping us to understand the physiological behaviour of muscle in terms of physical and biochemical properties of its constituent molecules, and has revealed mechanisms which seem remarkably simple, efficient and elegant. An enormous amount of work still remains to be done however, especially to describe these mechanisms in really detailed structural and chemical terms. And I think there are sufficient mysteries about muscle itself and about the use of the muscle mechanism in other cellular motile systems, to make this work continuously fascinating.

References

Andersson-Cedergren, E.: J. Ultrastruct. Res. Suppl. 1 (1959).
Ebashi, S., Endo, M.: Progr. Biophys. Mol. Biol. 5, 123 (1968).
Elliott, G.F., Lowy, J., Millman, B.M.: J. Mol. Biol. 25, 31 (1967).
Franzini-Armstrong, C., Porter, K.R.: Nature 202, 355 (1964).
Hanson, J., Huxley, H.E.: Symp. Soc. Exptl. Biol. 9, 228 (1955).
Hanson, J., Lowy, J.: J. Mol. Biol. 6, 46 (1963).
Haselgrove, J.C.: Cold Spring Harbor Symp. Quant. Biol. 37, 341 (1972).
Haselgrove, J.C., Huxley, H.E.: J. Mol. Biol. 77, 549 (1973).
Hill, A.V.: Proc. Roy. Soc. B135, 446 (1948).
Hill, A.V.: Proc. Roy. Soc. B136, 399 (1949).
Hitchcock, S.E., Huxley, H.E., Szent-Gyorgyi, A.G.: J. Mol. Biol. 80, 825 (1973).
Huxley, A.F.: Proc. Roy. Soc. B178, 1 (1971).
Huxley, A.F., Niedergerke, R.: Nature 173, 971 (1953).
Huxley, A.F., Simmons, R.M.: Nature 233, 533 (1971).
Huxley, A.F., Simmons, R.M.: Cold Spring Harbor Quant. Biol. 37 (1972).
Huxley, H.E.: Ph.D. Thesis, University of Cambridge (1952).
Huxley, H.E.: Biochim. Biophys. Acta 12, 387 (1953a).
Huxley, H.E.: Proc. Roy. Soc. B141, 59 (1953b).
Huxley, H.E.: J. Biophys. Biochem. Cytol. 3, 631 (1957).
Huxley, H.E.: J. Mol. Biol. 7, 281 (1963).
Huxley, H.E.: J. Mol. Biol. 37, 507 (1968).
Huxley, H.E.: Science 164, 1356 (1969).
Huxley, H.E.: Abstracts, 8th Int. Cong. Biochem. Interlaken (1970).
Huxley, H.E.: Abstracts, Amer. Biophys. Soc. p. 235(a) (1971a).
Huxley, H.E.: Biochem. J. 125, 85 (1971b).

Huxley, H.E.: Cold Spring Harbor Symp. Quant. Biol. 37, 361 (1972).
Huxley, H.E.: Nature 243, 445 (1973).
Huxley, H.E., Brown, W.: J. Mol. Biol. 30, 383 (1967).
Huxley, H.E., Hanson, J.: Nature 173, 973 (1954).
Huxley, H.E., Hanson, J.: Proc. 1st Europ. Regional Conf. Elect. Micro.
   Stockholm, p. 260 (1956).
Kendrick-Jones, J., Lehman, W., Szent-Gyorgyi, A.G.: J. Mol. Biol. 54,
   313-326 (1970).
Lehman, W., Kendrick-Jones, J., Szent-Gyorgyi, A.G.: Cold Spring
   Harbor Symp. Quant. Biol. 37, 319 (1972).
Lowey, S., Slater, H.S., Weeds, A.G., Bakev, H.: J. Mol. Biol. 42, 1
   (1969).
Lymn, R.W., Huxley, H.E.: in preparation (1975).
Lymn, R.W., Taylor, E.W.: Biochemistry 9, 2975 (1970).
MacLachlan, A.D., Stewart, M., Smillie, L.B.: J. Mol. Biol. (in press).
Margossian, S.S., Cohen, C.: J. Mol. Biol. 81, 409 (1973).
Moore, P.B., Huxley, H.E., Derosier, D.J.: J. Mol. Biol. 50, 279 (1970).
Ohtsuki, I., Masaki, T., Nonamura, Y., Ebashi, S.: J. Biochem. 61, 817
   (1967).
Parry, D.A.D., Squire, T.M.: J. Mol. Biol. 75, 33 (1973).
Pepe, F.: J. Cell Biol. 28, 505 (1966).
Porter, K.R., Palade, G.E.: J. Biophys. Biochem. Cytol. 3, 269 (1957).
Potter, J.D., Gergely, J.: Biochem. 13, 2697 (1974).
Reedy, M.K., Holmes, K.C., Tregear, R.T.: Nature 207, 1276 (1965).
Sodek, J., Hodges, R.S., Smillie, L.B.: Proc. Nat. Acad. Sci. USA 69,
   3800 (1972).
Spudich, J.A., Huxley, H.E., Finch, J.T.: J. Mol. Biol. 72, 619 (1972).
Squire, J.M.: J. Mol. Biol. 72, 291 (1973).
Vibert, P.J., Haselgrove, J.C., Lowy, J., Poulsen, F.R.: Nature New
   Biol. 236, 182 (1972).
Vibert, P.J., Lowy, J., Haselgrove, J.C., Poulsen, F.R.: Abstracts,
   1st European Biophys. Cong. Vienna (1971).
Wakabayashi, T., Huxley, H.E., Amos, L., Klug, A.: J. Mol. Biol.,
   in press (1975).
Weber, A., Bremel, D.R.: In 'Contractility of Muscle Cells and Related
   Processes' Ed. R.J. Podolsky. Englewood Cliffs, N.J.: Prentice-Hall,
   p. 34 (1972).
Weber, A., Murray, J.M.: Physiol. Rev. 53, 612 (1973).

Discussion

Dr. Rüegg: Does the change of intensity drop of the 143 layer line parallel the time course of stiffness changes, i.e. the time course of the active state?

Dr. Huxley: I am not completely familiar with the most recent mechanical measurements during the onset of the active state. Approximately it does - but certainly the accuracy of the measurements is not good enough. The error is ca. ± 5% calculated from an average of a dozen muscles. I am cautious of doing close comparisons with measurements of other time courses at all at the moment. In general one could say that the intensity change is ahead of the tension changes. Until we can do the measurements more rapidly it will be rather difficult to make this comparison with other measurements of the active state more precise.

Dr. Chaplain: As shown in the slide (see Fig.) an inital change in the pitch of the myosin helix is observed in oscillatory contraction-relaxation cycles of glycerinated frog sartorius muscle in presence

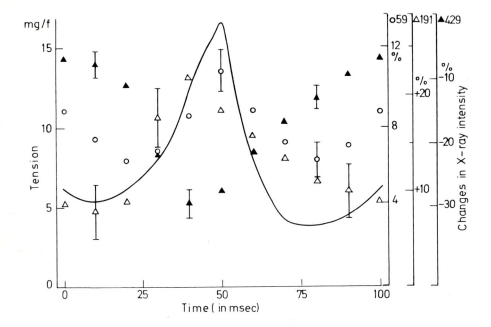

Time course of the changes in the pitch of the myosin helix and in
cross-bridge attachment during ascillatory contraction-relaxation
cycles of frog sartorius muscle. The tension record (in mg/fibre) is
drawn as a continuous line, while the intensity changes are indicated
by the respective symbols for 429 Å (▲), 191 Å (Δ) and 95 Å (o). The
error bars on the data provide some information on the characteristic
standard error at each of the three layer-lines

of 10 mM ATP and 5×10^{-7} M Ca^{2+}. The reciprocal intensity variations
in the 429 Å and 191 Å reflections definitely precede the cross-bridge
attachment by about 10 msec as measured by the intensity of the 59 Å
reflections. Would you think it is possible that changes in the cross-
bridge stalk (for example a shortening of the hydrogen bridges which
could induce a twisting of the double helix) may be the initial step
in energy transformation?

The energy of ATP hydrolysis could be stored as torsional energy
which is then transformed into a shearing force as the result of the
re-extension of the α-helical hydrogen bridges when the cross-bridge
attaches?

Dr. Wilkie: What were the conditions used for the experiments shown
in your slide, Dr. Chaplain?

Dr. Chaplain: Bundles of 15 glycerinated frog sartorius fibres were
attached to two glass rods, with one end connected to a vibrator, the
other to a tension transducer. The oscillatory length changes applied
through the vibrator were followed by delayed sinusoidal tension
changes, i.e. the fibres were doing work on the mechanical apparatus.
The fibre bundle, immersed in a Ca^{2+}-containing activating solution,
was sandwiched between two Mylar windows to allow the X-ray beam
from a Synchrotron to pass through (Chaplain and Sacharjan, FEBS
Letters 41, 50-32, 1974). While the total intensity of the off-merio-
donal 429 Å reflection was recorded, of the density distributed along

the 59 Å layer-line only the density near the meridian up to an equatorial spacing of 0.005 Å^{-1} was sampled. The intensity of the 191 Å layer-line was recorded with a position-sensitive counter at the radial spacing of 0.01–0.013 Å^{-1}. The total counting period for accumulating counts on each of the three reflections was 20 min at any interval given.

Dr. Wilkie: The experiment of Dr. Chaplain was different from that reported by H.E. Huxley. Changes following an electrical stimulus need some time for Ca^{2+} release whereas in Dr. Chaplain experiments Ca^{2+} was continually present. Could H.E. Huxley comment on this point?

Dr. Huxley: I am taking a very cautious and conservative position on this subject. What we are measuring is cross-bridge movement but we have very little evidence about the exact form of this movement at the moment. This movement is taking place with the right kind of rate for the model currently used. It involves a lot of hard work to measure these rate-constants accurately and see if they make sense. At present we are only able to do this on the two main sets of reflections, i.e. for the 143 meridonal and the equatorial one and gradually we shall move on to the other reflections. At the moment all that we can say is that these changes are taking place fast enough to correspond to cross-bridge attachments, but we cannot prove that these changes all arise from specific kinds of cross-bridge movements.

An Investigation of the Cross-bridge Cycle Using ATP Analogues and Low-angle X-ray Diffraction from Glycerinated Fibres of Insect Flight Muscle

K. C. Holmes, R. S. Goody, H. G. Mannherz, J. Barrington Leigh, and G. Rosenbaum

A. Introduction

I. Actomyosin

The interdigitating filament arrays of muscle contain as their major components the proteins actin and myosin. Globular actin molecules aggregate to form the "thin" filaments. Myosin molecules consist of fully α-helical tails and enzymatically active heads. The molecules spontaneously aggregate to form bipolar filaments (the "thick" filaments) with the heads protruding to form the "cross-bridges". Two heads may constitute one cross-bridge. The interaction between actin and myosin is mediated by the cross-bridges. By limited proteolytic digestion the enzymatically active fragments of myosin S_1 (single heads) and HMM (two heads with a length of tail) may be prepared. The heads (S_1) are joined to the tails by a more flexible part of the tail known as the S_2 region. (For details and references see Huxley, this volume.)

The complex between actin and myosin, namely actomyosin, catalyses efficiently the hydrolysis of MgATP. This reaction leads to the production of mechanical work in muscle. A second effect of ATP is to bring about the rapid dissociation of actomyosin into actin and myosin. Measurements of the rate of dissociation of actomyosin by ATP at physiological concentrations (Lymn and Taylor, 1971) show the dissociation step to be fast compared to the rate of cleavage of ATP, so that myosin alone is thought to be responsible for ATP cleavage in muscle.

Pure myosin (or HMM or S_1) is an ATPase but hydrolyses MgATP much more slowly than actomyosin. The release of the products of hydrolysis is rate-limiting (Taylor et al., 1970; Lymn and Taylor, 1970) the role of actin being to accelerate the release of products on reassociating with the myosin-product-complex. Thus under steady-state conditions a myosin solution in the presence of an excess of MgATP should contain as the major species myosin to which the products of hydrolysis (ADP, P_i; $P_i \equiv$ phosphate) are bound.

Detailed kinetic studies of the mechanism of hydrolysis of ATP by myosin by Trentham and collaborators (Bagshaw et al., 1974) show the binding of ATP and release of products to be two-step processes involving conformational changes (i.e. MATP $\rightleftharpoons$ M*ATP) after the ATP is bound. The conformational change on binding ATP manifests itself though an increase of the protein fluorescence. A second conformational change (M**) is prossibly associated with the cleavage step.

II. The Ca^{++} Control System

The activity of muscle is regulated by the level Ca^{++} ion in the sarcoplasm (see review by Weber and Murray, 1973 and articles by Perry et

al. in this volume). Natural thin filaments contain, in addition to
actin, troponin and tropomyosin. In the absence of Ca^{++} the affinity
of the myosin-ADP-P_i complex for actin is low because tropomyosin
molecules block the sites of attachment. On binding Ca^{++} to troponin,
the tropomyosin is moved so that myosin-ADP-P_i can bind to the thin
filaments leading to the release of the products of hydrolysis and
to the initiation of the conformational changes in the myosin-actin
complex that produce mechanical work. Some invertebrate muscles also
contain a Ca^{++} regulatory protein attached to myosin (see Kendrick-
Jones, this volume).

III. Glycerinated Fibres as a Model of Muscular Contraction

The treatment of muscle fibres with 50% glycerol at $-20^{O}C$ destroys the
membranes and releases the soluble proteins of the sarcoplasm. Glycerinated fibres from rabbit psoas will contract when MgATP is added
under the correct ionic conditions (Szent-Györgyi, 1949). Thus glycerinated preparations may be used to examine the effects of chemical
modifications to substrate or protein. Glycerinated fibres from the
longitudinal flight muscles of the giant water bug *Lethocerus maximus*
will generate work continuously in an oscillatory manner in the presence of ATP, Mg^{++} and Ca^{++} (Jewell et al., 1964). These muscle fibres
also give rise to a detailed X-ray low-angle diffraction pattern which
makes them very suitable for our present studies.

When glycerinated fibres are transferred to a neutral buffer solution
not containing ATP, the fibres take up the rigor state which is characterised by high stiffness reflecting a high degree of interaction
between actin and myosin.

In the presence of 5 mM Mg ATP and EGTA ($[Ca^{++}] < 10^{-7}$ M) the stiffness
drops to a low value. The interaction between actin and myosin is now
small. The myosin steady-state complex consists of myosin-ADP-P_i (Marston and Tregear, 1972) which, in the absence of Ca^{++}, has low affinity
for natural thin filaments. This state, the relaxed state, emulates
the living resting state.

One of the working hypotheses of current muscle research is that these
two states, rigor and relaxed, represent or are similar to states
present transiently in actively contracting muscle.

B. The Structure of the Rigor and Relaxed States in Insect Flight Muscle

I. Rigor

Low-angle X-ray diffraction patterns from glycerinated longitudinal
flight muscle fibres from *Lethocerus maximus* show a characteristic
crystalline diffraction pattern (Fig. 1) (Reedy et al., 1965; Miller
and Tregear, 1972). This diagram was partially interpreted by Reedy
et al. with the aid of electron micrographs of thin longitudinal sections. The X-ray scattering arises from the cross-bridges. In rigor
the cross-bridges make an angle of about 45° to the filament axis and
are firmly attached to the actin filaments. The average repeat of the
pattern of attachment in any one electron microscope section is 380 Å
(this corresponds to the 388 Å repeat seen by X-ray diffraction). However, the true repeat is 1,160 Å (Reedy, 1968).

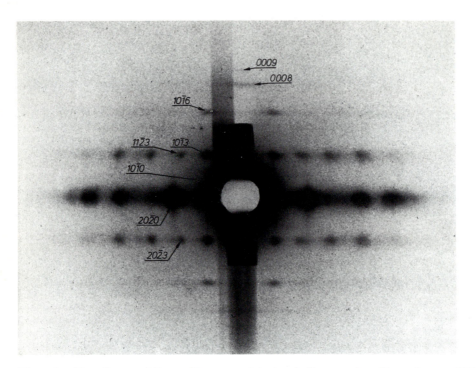

Fig. 1. The X-ray fibre diagram obtained from a bundle of ca. 100 glycerinated insect flight muscle fibres (*Lethocerus maximus*) irrigated in a flow cell with "rigor" solution (no ATP present) which additionally contains 20 mM $MgCl_2$, EGTA to complex traces of Ca^{++} and buffer, pH 6.9. X-ray source: DESY-electron synchrotron (Hamburg), 7.2 GeV, 9 mA. Focusing system: bent fused-silica mirror and bent quartz monochromator (10$\bar{1}$1 plane), λ = 1.5 Å; specimen-film distance 1,5 m, exposure time 1 h using Ilford Industrial G film. The hexagonal indexing is shown a = 520 Å, c = 1,160 Å. (Goody et al., 1975)

The lattice symmetry is hexagonal with a cell constant of about 520-560 Å. The myosin filaments sit at the corners of the hexagonal cell with apparent six-fold symmetry. The actin filaments sit on the two-fold positions. The space group at low resolution is P6$_4$ with a 388 Å c-axis pseudo-repeat (The strongest evidence that it is P6$_4$ comes from the electron microscope studies of Reedy, 1968. From X-rays alone it would not be possible to decide between the similar space groups P3$_1$, P3$_2$, P6$_2$, P6$_4$). The existence of a three-fold screw axis (which defines the four possible space groups for the small pseudo-symmetric cell) is indicated by the absence of meridional reflexions on layer lines 3 and 6 and the presence of a meridional reflexion on layer line 9 (these are layer lines 1,2 and 3 of the pseudo-cell). At wider angles of diffraction other layer lines are seen which do not belong to the 388 Å repeat and the symmetry drops to P6. Thus the P6$_4$ symmetry is only a pseudo-symmetry. The most important of the reflexions breaking pseudo-symmetry is the weak 145 Å meridional reflexion (0008). The characteristic X-ray diffraction features are: the two noteworthy equatorial reflexions (10$\bar{1}$0, 20$\bar{2}$0) are equally strong (Miller and Tregear, 1970); the innermost reflexion on the 388 Å layer line (10$\bar{1}$3) is strong; the corresponding reflexion on the 194 Å layer line (10$\bar{1}$6) is weak; the 145 Å meridional (0008) and 129 Å meridional (0009) are weak.

An interpretation of the X-ray diffraction pattern may be based upon the following postulates:

1. The low-angle scattering arises chiefly from the cross-bridges.

2. The $P6_4$ pseudo-symmetry is given only by <u>attached</u> cross-bridges and is a manifestation of actomyosin interaction.

3. The azimuthal angular relationship between myosin and actin is the major factor determing actomyosin interaction. Thus the actin filament symmetry largely determines the attachment sites since the actin filaments rotate so that an equivalent (or nearly equivalent) binding site sees each myosin filament every 388 Å. (At low resolution the actin filament has pseudo two-fold symmetry.) This fact alone accounts for the 388 Å repeat of the pseudo-symmetric $P6_4$ unit cell.

4. The cross-bridges originate from the myosin filament with an axial spacing of 145 Å and a symmetry which is effectively hexagonal (Squire, 1972). This is the second factor affecting acto-myosin interaction in the lattice. In attaching to actin, the outer ends of the cross-bridges are deformed so that they no longer conform to the 145 Å repeat of the myosin filament. Thus the $P6_4$ pseudo-symmetry reflexions may be thought of as "ghost" reflexions, being a manifestation of the distortions of the myosin helix symmetry introduced by binding to actin.

Using these postulates the arrangement of attached cross-bridges shown in Fig. 2 may be generated. The correspondence between this and the transverse thin sections obtained by Reedy (1968, particularly plate V) is very good, which is not altogether surprising since our postulates are closely based upon Reedy's arguments and results.

One striking feature of Reedy's micrographs, namely "double chevrons", is not accounted for. However, a straightforward extension of our argument allows this to be included. It is first convenient to define as θ the angle between the diameter of the myosin molecule pointing at the actin molecule (a cell edge) and the corresponding diameter of actin at the same level. Thus in the cases of attached bridges shown in Fig. 2 θ is zero. For the neighbouring actin molecule $\theta = 60°$. If we now postulate that the probability of attachment is a periodic function $P_\theta(\theta)$ with an effective width of about $60°$ then a second site of attachment at the same level (broken line bridges) becomes possible. This situation occurs frequently in Reedy's transverse sections which we take as *a priori* evidence that the angular component of the probability of cross-bridge attachment has such a form. By combining the two sites at level 1 with two sites related by the screw triad at level 2 the typical double chevrons seen in longitudinal section are generated (Fig. 2b).

A second factor influencing cross-bridge interaction is the vertical separation of cross-bridges. It is clear that the cross-bridges are quite flexible in the Z-direction so that the interaction probability can also be written as a periodic function $[P_z(Z)]$. By combining P_z and P_θ with suitable parameters it seems likely that most of the arrangements of cross-bridges seen by Reedy can be generated.

The existence of the double chevron also gives a simple explanation of the strong 10$\bar{1}$3. The trace of these Bragg planes through the structure is shown in Fig. 2b and it is clear that the double chevron leads to a concentration of mass on these planes. One can see that this depends on two factors: the degree of angling of the bridges and the degree of occupancy of the third site. If, for example, the third (empty) site were filled then one would expect the 10$\bar{1}$3 reflexion to

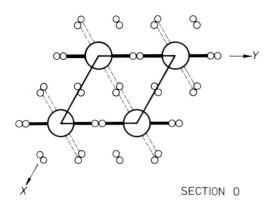

SECTION 0

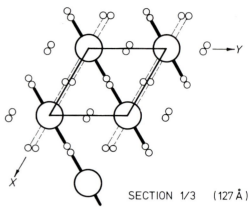

SECTION 1/3   (127 Å)

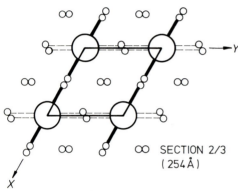

SECTION 2/3
(254 Å)

Fig. 2a. The arrangement of bound
cross-bridges in insect flight
muscle in rigor shown in transverse
sections. The myosin (thick) filaments (large circles) sit at the
corners (6-fold positions) of a
hexagonal lattice. The actin (thin)
filaments (pairs of small circles)
sit on the two-fold positions. As
the actin filaments pass up through
the unit cell they rotate anti-
clockwise so that each strand is
$180°$ away from the initial orientation after 388 Å. If one postulates that whenever an actin fila-
ment faces a myosin filament a bridge binds, the pattern of bound
bridges shown with heavy lines is produced. This arrangement of bridges
has $P6_4$ space group symmetry giving layers of bridges each $60°$ away
from its neighbours. If, in addition, an actin filament disorientated
by up to $+60°$ from the ideal position is allowed to bind a bridge, then
the second set of bridges (broken lines) is also produced. The third
set of actin disorientated by $-60°$ from the ideal orientation apparently do not bind

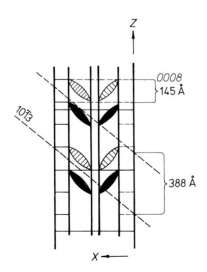

Fig. 2b. A longitudinal section containing
the x and z axes. The second-site bridges
are shown cross-hatched. The cross-bridges
originate every 145 Å along the myosin
filament but bind to the thin filaments
with periodicity mostly determined by the
388 Å repeat of actin. The $45°$ angle of
the bridge was first reported by Reedy
et al. (1965). Note how the first- and
second-site bridges form the characteristic "double chevron" (Reedy, 1968).
The traces of the $10\bar{1}3$ and 0008 Bragg
planes are also shown. The distribution
of density shown would strengthen the
$10\bar{1}3$ and weaken the 0008 reflexions

become weak. This explanation may be important in understanding the mode of action of one of the analogues.

II. Relaxed

In the presence of 5 mM Mg ATP and EGTA ($[Ca^{++}] < 10^{-7}$ M) the stiffness of the fibres drops to 1/10 of the rigor value. Under these conditions, actin and myosin have a low affinity for each other. This state, the relaxed state, emulates living resting muscle. Its X-ray

Fig. 3. X-ray fibre diagram (details as in Fig. 1) from glycerinated insect muscle fibres irrigated with a "rigor" solution containing additionally 15 mM ATP. Exposure time 2 h. (Goody et al., 1975). Note the strong meridional 0008 reflexion

diffraction pattern (Fig. 3) is quite different from and much simpler than that of rigor. Only traces of the $P6_4$ lattice reflexions remain. The 145 Å meridional reflexion (0008 on the rigor lattice indexing) is strong and narrow in the equatorial direction; its second order (72 Å) is relatively strong. On the equator the $10\bar{1}0$ is stronger than the $20\bar{2}0$. The diffration pattern arises from the cross-bridges, spaced every 145 Å along the myosin filament, projecting at right angles to the myosin filament (see Fig. 6, Reedy et al., 1965). The narrowness of the 145 Å reflection shows that the myosin filaments are in longitudinal register in relaxed muscle. The change in the ratio of the equatorial reflection ($10\bar{1}0/20\bar{2}0$) on going from the relaxed to rigor state shows that a considerable transfer of mass from myosin to actin takes place on going into rigor (Miller and Tregear, 1970; Huxley, 1968).

From Lymn and Taylor's kinetic scheme for the hydrolysis of ATP by actomyosin one would expect muscle fibres saturated with MgATP to contain myosinADP.P_i as the major species. Chemical analysis of the fibres shows this to be the case (Marston and Tregear, 1972). Thus we can associate the biochemical state myosinADP.P_i with the right-angled configuration of the cross-bridges. If the kinetic constants of rabbit myosin can be extrapolated to insect, then the steady-state complex would be M**ADP.P$_i$ in the Trentham scheme (see Bagshaw et al., 1974). This point will be discussed further below.

C. The Cross-bridge Cycle

On the basis of observations of the changes in the X-ray diffraction pattern of frog muscle when it is activated Huxley (1969) (see also this volume) proposed a model for the motion of the cross-bridges during muscular contraction whereby the cross-bridges "row" the actin filaments past the myosin. Such a model also suggests itself from the "in vitro" studies discussed above (Reedy et al., 1965; Pringle, 1967). By combining the Huxley rowing hypothesis with the results of their kinetic studies of actomyosin, Lymn and Taylor (1971) formulated the cross-bridge cycle shown in Fig. 4. The binding of ATP dissociates

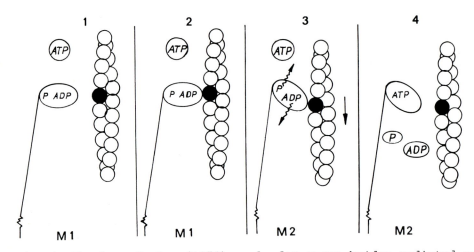

Fig. 4. The Lymn-Taylor (1971) cycle for cross-bridge mediated contraction. This is based upon their own kinetic studies and the structural studies of activated contraction muscle of H.E. Huxley (1969). The existence of "up" and "down" states is postulated on the basis of the studies of rigor and relaxed insect muscle (see text). The cross-bridge in the "down" position binds ATP which causes it to dissociate from actin. The ATP is hydrolysed to ADP and phosphate (P$_i$) and the cross-bridge goes to the "up" position. It reassociates with actin which promotes release of the products of hydrolysis and at about the same time undergoes a conformational change to the "down" position while bound to myosin. The nucleotide binding site is now free to bind a further ATP-molecule

the cross-bridges which then return to the right-angled configuration. At some stage, ATP is cleaved but the products of hydrolysis stay attached to the cross-bridge. The myosinADP.P$_i$ complex has apparently a higher affinity for actin than myosinATP so that it reassociates with actin bringing about release of products. At the same time a change in the angle between the cross-bridge and actin occurs which

leads to the movement of actin past myosin (the power stroke). At the
bottom of the power stroke ATP binds and the cycle starts again.

A number of questions arise concerning the details of this cycle. For
example, at which moment during the power stroke are the products of
hydrolysis released, or how many distinct conformations of myosin
must be postulated?

A further question concerns the nature of the conformational change of
the cross-bridge when attached to actin: does it roll on the actin
without the actin or cross-bridge undergoing a conformational change,
as suggested by A.F. Huxley and Simmons (1971), or does the cross-
bridge bind to actin at a fixed angle and undergo an extensive conformational change? A related question is whether the "back stroke"
of the cross-bridge is "driven" by ATP hydrolysis or is relatively
passive. We earlier thought that we had evidence to show that the back
stroke was driven (Mannherz et al., 1972) but our recent results (Goody
et al., 1975) have obliged us to abandon this contention.

In order to illuminate some of these problems, we have been investi-
gating the effects of three ATP analogues on the low-angle X-ray diffraction patterns of glycerinated insect flight muscle.

D. ATP Analogues

We have studied three analogues of ATP which have modifications to
the tripolyphosphate moity: ATP(α,β-CH$_2$) Miles Biochemical, ATP(γ-S)
(Goody and Eckstein, 1971), and ATP(β,γ-NH) (Yount et al., 1971). All
three analogues bind to myosin and are either substrates or compet-
itive inhibitors of myosin ATPase.

The first two analogues are hydrolysed by rabbit myosin but with rate
constants very different from those of ATP (see Goody et al., 1975). In
each case the cleavage step is slow but the release of products is
fast and not accelerated by actin. Consequently the turnover of analogue by myosin is higher than for ATP even though the cleavage-step
is slow. However, at high levels of analogue concentration the predominant steady-state complex should be myosinATP-like. More specifically, it should be M*ATP-like.

Because the hydrolysis product of ATP(γ-S) is ADP, which may be re-
phosphorylated to ATP by adenylate kinase in the muscle fibres, precautions have to be taken to eliminate ATP when using this analogue.
A-P$_5$-A is added to inhibit the adenylate kinase (Lienhard and Secemski,
1973). An acid phosphatase is added to hydrolyse the ATP and ADP produced.

Since the turnover of both analogues is relatively high, the fibre
bundle should be thin (not more than 50 fibres) and the analogue
should be continuously replenished by flowing the solution across the
muscle fibres in order to keep a saturating concentration of analogue
in the fibres.

ATP(β,γ-NH) is not hydrolysed by myosin. The nature of the complex
formed with myosin is apparently ADP-like as was first suggested by
Barrington Leigh et al. (1972). This point is discussed below.

E. The Relaxed State May be Achieved without Cleavage

The stiffness of insect flight muscle fibres immersed in a buffered solution (pH 6.9) containing Mg ATP(γ-S) or Mg ATP(α,β-CH$_2$) falls to

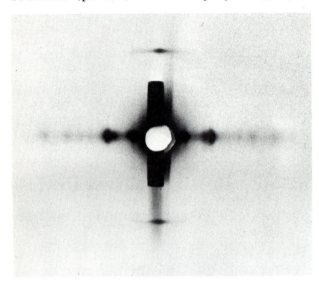

Fig. 5. X-ray fibre diagram (details as in Fig. 1) from glycerinated insect muscle fibres irrigated with "rigor" solution containing additionally 20 mM ATP(γ-S) at 4°C. Possible regeneration of ATP by adenylate kinase was avoided by adding AP$_5$A and acid phosphatase. Note the similarity of this diagram to Fig. 3

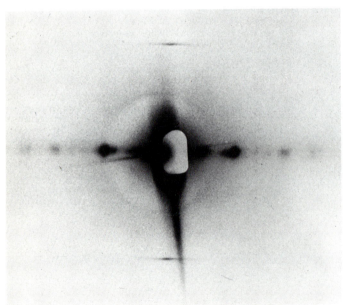

Fig. 6. X-ray fibre diagram from a small fibre bundle (ca. 40 fibres) in a flow cell irrigated with a "rigor" solution containing additionally 30 mM ATP(α,β-CH$_2$) at 2°C, pH 6.9 (18-inch copper rotating anode X-ray generator, mirror-monochromator focusing system CuKα λ = 1.54 Å, specimen film distance 1.2 m, exposure time (Ilford G film) 12 h). This pattern is essentially identical to a relaxed diagram (Fig. 3). (Goody et al., 1975). In addition traces of the 194 Å layer line can be seen

the Mg ATP value for analogue concentrations of 20 mM and 30 mM respectively. Figs. 5 and 6 show the X-ray diagrams obtained from ATP (γ-S) and ATP(α,β-CH$_2$) at these concentrations using small fibre bundles, a flow cell, and precautions to eliminate ATP as outlined in the previous section.

The X-ray diagrams are essentially identical to those obtained with ATP. Since the steady-state complex obtained with ATP is M**ADP.P$_i$ we infer that at the level of resolution of the low-angle X-ray diffraction ($\simeq$ 50 Å) the states M*ATP and M**ADP.P$_i$ have to be related to the same cross-bridge configuration.

F. An Intermediate State Produced by Low Concentrations of ATP(β,γ-NH)

I. X-ray Characteristics

The X-ray diffraction pattern produced by insect flight muscle fibres in 1 mM Mg ATP(β,γ-NH) at pH 6.9, 4°C is shown in Fig. 7. It differs

Fig. 7. X-ray fibre diagram; details as in Fig. 6 except that the muscle was irrigated with a "rigor" solution containing in addition 1 mM ATP (β,γ-NH) 4°C. Note in particular the well-developed low-angle layer lines, the strong 0008, the strong 10$\bar{1}$6 and weak 10$\bar{1}$3

from rigor in the following ways: the 10$\bar{1}$0/20$\bar{2}$0 ratio is intermediate (1.7) between rigor and relaxed; the 10$\bar{1}$6 is stronger than the 10$\bar{1}$3; the 20$\bar{2}$3 is stronger than the 11$\bar{2}$3; the 0008 (145 Å meridional) is much broader than in relaxed muscle. On warming the muscle to 23°C the intensity changes tend to revert towards rigor (Goody et al., 1975). On addition of analogue a fairly large change in the spacing of the hexagonal unit cell of the muscle takes place. The cell becomes smaller. We generally observed 6% change and on one occasion observed a 12% change. The change in ionic strength produced by adding analogue is minimal and cannot be invoked to explain this effect.

II. Mechanical Characteristics

On adding 1 mM Mg ATP(β,γ-NH) to fibres under tension the tension
drops without much change in the stiffness (Barrington Leigh et al.,
1972). This indicates that the fibres have lengthened. This phenomena
has been studied by Kuhn (1973) who finds that fibres lengthen and
shorten reversibly on addition and removal of MgATP(β,γ-NH). The amplitude of the induced length change is small (0.5%) and could correspond
to a change in the angle of the cross-bridge.

III. The Nature of the Intermediate State

The first question is whether the new state is not a trivial mixture
of the rigor and relaxed states. Although the main equatorial reflexions could be accounted for by a linear superposition of the relaxed
and rigor diffraction patterns other regions of the diffraction pattern cannot be explained in this way: two important low order reflexions become stronger in absolute terms, and two become weaker.
Furthermore, the radial extent of the 0008 is much greater in the
presence of analogue than in relaxed fibres. These changes in the diffraction pattern cannot be explained as a linear superposition of
rigor and relaxed. Lastly, the change of hexagonal spacing observed
makes this kind of explanation unlikely.

The intermediate X-ray diagram can most readily be interpreted by
assuming that the cross-bridges are in a relaxed-like configuration
but are attached to the actin filaments ("up" and "on") (Fig. 8).
This explanation was first advanced by Barrington Leigh et al. (1972)

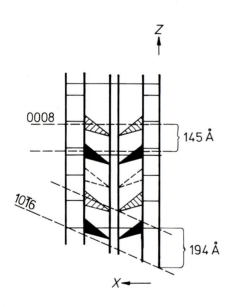

Fig. 8. A longitudinal section
through the "intermediate" arrangement of cross-bridges produced by
binding ATP(β,γ-NH). The major
features of the structure may be
explained in terms of the P6₄ symmetry of the actin binding sites
as in rigor (Fig. 2). However, in
addition to the "double chevron"
binding sites characteristic of
rigor (see Fig. 2b) the third site
(broken line) may also be partially
filled. The cross-bridge no longer
appears to have the strongly-angled
"down" form characteristic of rigor
but is shorter and stumpier with
more mass near the myosin filament.
The traces of the 0008 and 101̄6
Bragg planes are shown. It can be
seen that the proposed shape and
distribution of cross-bridges would
strengthen these Bragg reflexions
(cf. Fig. 2b)

and has been more forcefully proposed by Goody et al. (1975). The
strong 145 Å meridional reflection is the strongest indication that
a large proportion of the cross-bridges are "up" although this evidence,
taken alone, does not prove the bridges to be "on". The 145 Å reflection could have two components:

1. A proportion of the mass of the <u>attached</u> cross-bridges which lies near the myosin filament and which scatters according to the myosin repeat. This mass apparently varies according to whether the bridge is "up" or "down".

2. Free bridges unable to find actin binding sites which return to the relaxed (up-off) configuration on binding ATP(β,γ-NH).

The evidence that most bridges are "on" in the intermediate state comes from two sources:

1. The diffraction from the P6$_4$ symmetrical part of the structure (which is characteristic of acto-myosin interaction) is equally strong in the intermediate state and in rigor.

2. The mechanical stiffness is high.

By studying that part of the structure with P6$_4$ symmetry we study the attached cross-bridges. Thus the strengthening of the $10\bar{1}6$ and 0009 reflection is evidence for the <u>attached</u> cross-bridges being in a more nearly right-angled configuration even if the interpretation of the 145 Å reflection were ambiguous.

The weakening of the $10\bar{1}3$ on adding ATP(β,γ-NH) reflection is remarkable. It would be expected to be strong for a large range of cross-bridge shapes if the "double chevron" arrangement (Fig. 2) was maintained. The most straightforward explanation of its becoming weaker is to postulate that the third (empty) site becomes filled (dotted bridges Fig. 8). This would also lead to a strengthening of the 0008 (145 Å) and $10\bar{1}6$ (194 Å) as is indicated diagrammatically in Fig. 8. One infers from this that the probabilities of attachment P_θ at P_z must alter on binding ATP(β,γ-NH) so that binding in the intermediate state is <u>more probable</u> than in rigor.

The recent electron microscope studies by Beinbrech and Kuhn and by Tregear and Rodger, which have been privately communicated to us, show that the majority of cross-bridges lie at right angles in the presence of ATP(β,γ-NH). In particular the micrographs of Tregear and Rodger support the notion that bridges are attached regularly with 145 Å Z-periodicity.

The equatorial reflections $10\bar{1}0$ and $20\bar{2}0$ have an intermediate ratio in the intermediate state, indicating some transfer of mass from the actin filaments towards the myosin filaments. The bridge length must alter between rigor and the intermediate state. One way this could happen would be if the "tail" attaching the myosin head to the myosin filament (which may be the S$_2$ portion of myosin) rolled back onto the myosin molecule thereby producing a considerable transfer of mass towards the myosin filament.

Alternatively, the transfer of mass could come about through a radical change in the shape of the S$_1$ region. In rigor the indications (mostly from micrographs) are that the bridges are long cylinders (about 200 Å long). In relaxed muscle (and in the intermediate state) the shape of the bridge is probably more globular and is shorter ($<$ 150 Å long; Miller and Tregear, 1972). The analogue-induced change of side-to-side spacing (which amounts to 15-20 Å per bridge in the horizontal plane) would be in agreement with an active change of length of the bridge produced by binding ATP(β,γ-NH). Kuhn's (1973) measurements of an induced change in the Z-direction also support this idea.

G. The Possible Significance of the Intermediate State in the Cross-bridge Cycle

Barrington Leigh et al. (1972) suggested that ATP(β,γ-NH) might be an analogue of ADP because of the high affinity of myosinATP(β,γ-NH) for actin. Recently Tregear (private communication) has shown that ADP produces effects similar to those produced by ATP(β,γ-NH) on the diffraction pattern of *Lethocerus* flight muscles if care is taken to eliminate the resynthesis of ATP by inhibiting adenylate kinase. We have confirmed this result (to be published elsewhere). Thus it seems likely that ATP(β,γ-NH) at pH 6.9 is an analogue of ADP and not of ATP. If we take the intermediate state to typify the myosin-ADP state it is fairly straightforward to incorporate the analogue studies in an extended Lymn-Taylor cross-bridge cycle as is shown in Fig. 9. We have at the same time incorporated the more detailed information made available by Trentham's studies of the two-step nature of the binding and release of ATP and ADP. We are unable to equate the myosin conformations M*ATP and M*ADP as was suggested by the Trentham scheme since although they both have approximately the same conformation ("up"), the affinity of M*ADP for actin is high whereas the affinity of M*ATP for actin is low. To distinguish between the states of myosin with high and low affinity for actin we use the additional subscript r.

In summary, the binding of all the nucleotides we have tried produces the "up" state of myosin (M*ATP, M**ADP or M_r'ADP). However, the ability to dissociate acto-myosin varies widely between nucleotides and is particularly marked between ATP and ADP. In this respect ATP (β,γ-NH) seems to be an analogue of ADP and produces an intermediate state in which the bridges have an "up" orientation but retain high affinity for actin.

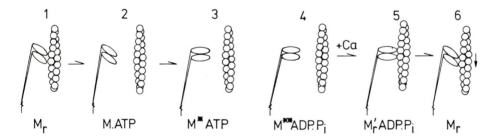

Fig. 9. An extended cross-bridge cycle (cf. Fig. 4) containing additionally the information derived from the solution studies of HMM and S_1 (Bagshaw et al., 1974) and the indicative deductions from our own studies of ATP analogues (Goody et al., 1975). The asterisks represent states with high protein fluorescence. We propose that binding ATP(β,γ-NH) to glycerinated muscle fibres imitates state 5. The possible relationship between the state M_r'ADP.P_i and states identified in solution remains to be characterised. The states M_r' and M_r (myosin in rigor) each have high affinity for actin. The binding of ATP(γ-S) or ATP(α,β-CH$_2$) imitates state 3. State 4 is produced by adding an excess of ATP. Low-angle X-ray diffractions from fibres in the presence of ATP, ATP(α,β-CH$_2$) and ATP(γ-S) all look very similar (Figs. 3, 5 and 6)

Acknowledgments. We are very grateful to Dr. Richard Tregear and
Prof. G. Beinbrech for discussing their unpublished work with us. We
thank H. Pfrang and M. Isakov for their excellent technical assistance.
We thank the staff of the EMBL outstation at DESY (Hamburg) for their
untiring efforts on our behalf.

References

Bagshaw, C.R., Eccleston, J.F., Eckstein, F., Goody, R.S., Gutfreund,
  H., Trentham, D.R.: Biochem. J. 141, 351-364 (1974).
Bagshaw, C.R., Trentham, D.R.: Biochem. J. 133, 331-349 (1973).
Barrington Leigh, J., Holmes, K.C., Mannherz, H.G., Rosenbaum, G.,
  Eckstein, F., Goody, R.: Cold Spring Harbor Symp. Quant. Biol. 37,
  443-447 (1972).
Caspar, D.L.D., Holmes, K.C.: J. Mol. Biol. 46, 99-133 (1969).
Goody, R.S., Eckstein, F.: J. Amer. Chem. Soc. 93, 6252 (1971).
Goody, R.S., Holmes, K.C., Mannherz, H.G., Barrington Leigh, J.,
  Rosenbaum, G.: Biophys. J., in press (1975).
Huxley, A.F., Simmons, R.M.: Nature 233, 533-538 (1971).
Huxley, H.E.: Science 164, 1356-1366 (1969).
Jewell, B.R., Pringle, J.W.S., Rüegg, J.C.: J. Physiol. 173, 6-8
  (1964).
Kuhn, H.J.: Sep. Experientia 29, 1086-1088 (1973).
Lienhard, G.E., Secemski, I.I.: J. Biol. Chem. 248, 1121-1122 (1973).
Lymn, R.W., Taylor, E.W.: Biochemistry 9, 2975-2983 (1970).
Lymn, R.W., Taylor, E.W.: Biochemistry 10, 4617-4624 (1971).
Mannherz, H.G., Barrington Leigh, J., Holmes, K.C., Rosenbaum, G.:
  Nature New Biology (Lond.) 241, 226-229 (1973).
Marston, S.B., Tregear, R.T.: Nature New Biology (Lond.) 235, 23-24
  (1972).
Miller, A., Tregear, R.T.: Nature (Lond.) 226, 1060-1061 (1970).
Miller, A., Tregear, R.T.: J. Mol. Biol. 70, 85-104 (1972).
Reedy, M.K.: J. Mol. Biol. 31, 155-176 (1968).
Reedy, M.K., Holmes, K.C., Tregear, R.T.: Nature (Lond.) 207, 1276-
  1280 (1965).
Squire, J.M.: J. Mol. Biol. 72, 125-138 (1972).
Szent-Györgyi, A.: Biol. Bull. 96, 140 (1949).
Taylor, E.W., Lymn, R.W., Moll, G.: Biochemistry 9, 2284-2991 (1970).
Weber, A., Murray, J.M.: Physiol. Rev. 53, 612-673 (1973).
Yount, R.G., Babcock, D., Ballantyre, W., Ojala, D.: Biochemistry 10,
  2484-2489 (1971).

Discussion

Dr. Hasselbach: Do the ATP analogs dissociate actomyosin?

Dr. Holmes: In our early work on the α,β-methylene ATP derivative,
investigation in the ultracentrifuge indicated a dissociation at relatively low concentration. Therefore I think the situation in the
fibre is different from that in solution. It could well be that you
get the same sort of effect with other analogs at much lower concentrations in solution.

Dr. Hasselbach: What is the effect of pyrophosphate?

Dr. Holmes: We have not investigated the effect of pyrophosphate. Some work has been done by the Oxford group, but somehow they have never published it.

Dr. Rüegg: Judging from electronmicroscopic studies, the effect of pyrophosphate on the arrangement of the cross-bridges seems to be similar to that of the AMP-PNP analog (see Beinbrech et al., Experientia 1972).

Dr. Hasselbach: Does the AMP-PNP analog or pyrophosphate produce dissociation of dissolved actomyosin?

Dr. Holmes: I think it is difficult to comment because the two substances produce an intermediate state which is difficult to interpret.

Dr. Beinbrech: We too obtained an intermediate state which is difficult to explain. If you look at electron micrographs of pyrophosphate or AMP-PNP-relaxed sarcomeres, cross-bridges seem to be in different positions: in some areas most of the cross-bridges seem to be angled, in other ones most of them seem to be relaxed. In agreement with these findings optical diffraction patterns seem to combine the patterns of rigor and ATP relaxed muscles. The positions of the cross-bridges of pyrophosphate and AMP-PNP-relaxed muscles seem to be influenced by the presence of phosphate. If phosphate buffer is used instead of imidazol the optical diffraction patterns of these relaxed fibres change into the direction of the rigor muscle. We think therefore that phosphate might influence the equilibrium postulated by Lymn and Huxley (Cold Spring Harbor Symp. Quant. Biol. 37, 449 (1972).

Dr. Holmes: I think my own feeling was that your results obtained with pyrophosphate do in fact reinforce our views and they do seem to fit.

Dr. Weber: Getting different answers from experiments in solution and with fibres does not mean necessarily that the findings contradict each other. In solution the protein concentrations are much lower than in the fibre. If the binding constant for the interaction of actin and myosin is higher in the presence of AMP-PNP than in the presence of ATP one may find that actomyosin in solution is dissociated by AMP-PNP but not at the high protein concentration in the fibre. If the AMP-PNP does not detach the bridge but makes it move from acute angles to a right angle, then you would expect a phenomenon like latency relaxation. If a fibre in rigor is under some tension to which the analog is then added, it should stretch and the tension should go down. Has that been found?

Dr. Rüegg: We added the AMP-PNP analog to fibres under tension in isometric experiments and found that they relaxed. If we removed the analog tension was regained. These data seem to fit in with the suggested model of an intermediate state of cross-bridges which are attached vertically to the thin filaments in the presence of the AMP-PNP. If this is so, it can be predicted that the immediate stiffness (dynamic stiffness) would be the same in the presence of a low concentration of the AMP-PNP analogue and in rigor, a prediction which has already been verified by Dr. Hans Kuhn, Herzig and myself when stiffness was measured by the application of quick stretches or releases (within one msec) of glycerinated *Lethocerus* muscle fibres.

Dr. Holmes: If a muscle has developed a so-called rigor-tension by washing out ATP or the analogs, the tension drops away. The tension can drop away in a number of ways: Either the stiffness of the bridges goes down, or the number of bridges bound goes down, or some intrinsic change in length takes place. The question is which of these phenomena is predominant. I think Dr. Kuhn has a much better answer to this problem and also I hear that the Oxford group has some answers, namely that there is an intrinsic change in length.

Dr. Huxley: What change of length would be expected on the basis of the tension regenerated when washing out the AMP-PNP analog?

Dr. Rüegg: If the 200 $\overset{\circ}{A}$ long cross-bridge moves once producing a displacement of 50-100 $\overset{\circ}{A}$, a change of maximally 0.5 to 1% is expected.

Dr. Holmes: I think it is less than that. It may be 0.3%.

Dr. Wilkie: Pyrophosphate apparently behaves like an ATP analog. Can we conclude that the phosphate end and not the adenine moiety of ATP attaches to the binding site?

Dr. Holmes: Yes, probably.

The Binding and Cleavage of ATP in the Myosin and Actomyosin ATPase Mechanisms

J. F. Eccleston, M. A. Geeves, D. R. Trentham, C. R. Bagshaw, and U. Mrwa

A. Foreword

One of H.H. Weber's great contributions to muscle biochemistry and physiology was his understanding of the essential role of ATP in muscle contraction, so clearly expressed in his 1954 review (Weber and Portzehl, 1954). It is therefore a great privilege for us to contribute a paper at this meeting on aspects of the mechanism of ATP hydrolysis catalysed by myosin and actomyosin.

B. Introduction

Studies of the myosin and actomyosin ATPases enable one to investigate how ATP hydrolysis occurs in relaxed muscle and in activated muscle that is shortening. The actomyosin ATPase activity is about three orders of magnitude greater than that of myosin, which is to be expected in view of the capacity of active muscle to do mechanical work.

One reason for the advances in our understanding of the kinetics of ATP hydrolysis over the past decade has been the availability of heavy meromyosin subfragment 1 prepared by mild proteolytic digestion of myosin (Lowey et al., 1969). Subfragment 1 exhibits ATPase activity, interacts with actin and, in contrast to myosin, does not aggregate at physiological ionic strength and so is more amenable to spectroscopic studies.

The mechanism of the Mg^{2+}-dependent myosin ATPase has been analysed using a variety of rapid reaction and isotope techniques. From these studies a seven-step scheme has been proposed [Eq. (1)] (Bagshaw et al., 1974):

$$M + ATP \overset{1}{\rightleftharpoons} M.ATP \overset{2}{\rightleftharpoons} M^*.ATP \overset{3}{\rightleftharpoons} M^{**}.ADP.P_i \overset{4}{\rightleftharpoons} M^*.ADP.P_i$$

$$\overset{5}{\rightleftharpoons} M^*.ADP + P_i \overset{6}{\rightleftharpoons} M.ADP \overset{7}{\rightleftharpoons} M + ADP \tag{1}$$

in which M represents subfragment 1 and k_{+i}, k_{-i} and K_i (= k_{+i}/k_{-i}) represent the forward and reverse rate constants and the equilibrium constant of the ith step. In this scheme the predominant steady state intermediate at $27^\circ C$, pH 8 and in 0.1 M KCl of the Mg^{2+}-dependent myosin subfragment 1 ATPase prepared from rabbit skeletal muscle is $M^{**}.ADP.P_i$ (Lymn and Taylor, 1970). The breakdown of $M^{**}.ADP.P_i$ is controlled by a process whose rate constant, $k_{+4} = 0.06s^{-1}$.

Features of the mechanism leading to the formation of M**.ADP.P_i are the high affinity of ATP for subfragment 1 (the overall association constant, K_1K_2, $\geqslant 10^9 M^{-1}$; Mannherz et al., 1974; Wolcott and Boyer, 1974), and the equilibrium constant of the cleavage step, K_3, which = 9 (Bagshaw and Trentham, 1973). The rate of formation of M**.ADP.P_i when molar excess of ATP is mixed with subfragment 1 is controlled either by the bimolecular association process with an apparent second order rate constant, K_1k_{+2}, = 1.8 × 10^6 $M^{-1}s^{-1}$, or at high ATP concentrations (>1 mM) by the M.ATP to M*.ATP isomerization with a first order rate constant, k_{+2}, = 400 s^{-1}(Bagshaw et al., 1974). However it is possible that k_{+3} is comparable with k_{+2} in which case both the isomerisation and the cleavage step control the rate of M**ADP.P_i formation at higher ATP concentrations.

Studies of oxygen incorporation into the products have made important contributions to our understanding of the ATPase mechanism. The oxygen atom introduced through hydrolysis is found in the product P_i. Other processes occur which give rise to the phenomenon of intermediate oxygen exchange (Levy and Koshland, 1959). This is the exchange of oxygen atoms during ATPase activity which results in the product P_i containing oxygen atoms from the solvent in addition to the one derived from hydrolysis. Sartorelli et al. (1966) showed that, if ATP was present in a large excess over the myosin and a substantial fraction of the ATP was hydrolysed in a medium containing $H_2^{18}O$, no incorporation of ^{18}O was detected in the remaining ATP in the solvent. This is expected in view of the tight binding of ATP to myosin. The ready reversibility of the cleavage step, M*.ATP $\rightleftharpoons$ M**.ADP.P_i, of Eq. (1) and the fact that M**.ADP.P_i is the predominant steady-state intermediate of the myosin ATPase suggest that intermediate oxygen exchange is occurring during the lifetime of M**.ADP.P_i and consequently M*.ATP is also likely to undergo oxygen exchange. If ATP labelled with ^{18}O in the 3 oxygen atoms of the γ-phosphate group (excluding the bridging oxygen) is mixed with a molar excess of subfragment 1 in $H_2^{16}O$ and the reaction is stopped at 2 s, the isolated ATP has lost about 75% of its label (Bagshaw et al., 1975). This ATP has originated from M*.ATP and so the oxygen exchange occurs as predicted. In the experiments we describe here, the phenomenon of intermediate oxygen exchange is used to learn more about the myosin and actomyosin ATPase mechanisms.

Lymn and Taylor (1971) have proposed a mechanism for the Mg^{2+}-dependent actomyosin ATPase [Eq. (2)], in which the predominant steady-state intermediate is also a myosin products complex. The greater the actin concentration, the more this complex will be associated with actin. The authors concluded that the major pathway for the actomyosin ATPase involved a step in which ATP cleavage occurred while myosin was dissociated from actin, since the dissociation rate of actin from the ternary ATP actomyosin complex is as fast or faster than the rate of transient protein—bound product formation. Eq. (2) describes their scheme.

$$AM + ATP \underset{1}{\rightleftharpoons} AM.ATP \underset{2 \atop -A}{\rightleftharpoons} M*.ATP \underset{3}{\rightleftharpoons} M**.ADP.P_i \underset{4 \atop +A}{\rightleftharpoons} AM.ADP.P_i$$

$$\underset{5}{\rightleftharpoons} AM + ADP + P_i \tag{2}$$

In Eq.(2) A represents actin and k'_{+i}, k'_{-i} and $K'_i (=k'_{+i}/k'_{-i})$ represent the forward and reverse rate constants and the equilibrium constant of the ith step. In order to test the validity of this scheme, it is pertinent to investigate whether the cleavage step is common to both myosin and actomyosin ATPase mechanisms. Asterisks have been introduced into intermediates of Eq. (2) to highlight the proposed similarity.

Two further reasons why it is important to answer this question are as follows. If the cleavage steps are found to be common to both ATP-ases, this provides support for Lymn and Taylor's ideas (1971) relating biochemical events to the events of the cross-bridge cycle. If Eq. (2) is correct, so that the cleavage step of the actomyosin ATPase is readily reversible, this has thermodynamic implications because it is likely that, when the cross-bridge is detached during the contractile cycle, the actual free energy change will be small (Bagshaw and Trentham, 1973).

The experiments which we describe here are designed to elucidate the details of the binding and cleavage of ATP in the myosin and actomyosin ATPases with particular emphasis on the relationship between the two mechanisms.

C. Results and Discussion

I. Kinetics of Intermediate Oxygen Exchange

As well as comparing the equilibrium constants of the cleavage step in the myosin and actomyosin ATPases, it is valuable to compare the rate constants associated with intermediate oxygen exchange. This may lead to an evaluation of the rate constants of the cleavage steps in the two ATPases. In the myosin ATPase it is not easy to evaluate k_{+3} and k_{-3} because preceding steps in the formation of $M^{**}.ADP.P_i$ are rate-limiting.

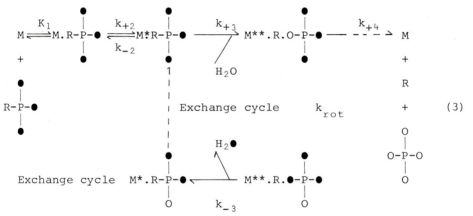

Eq. (3) describes a possible mechanism for intermediate exchange. O and ● differentiate oxygen atoms arising from H_2O and ATP respectively

R-P-●, R and O-P-O represent ATP, ADP and P_i respectively. k_{rot} is the rate of the process which occurs to allow oxygen atoms emanating from ATP to be eliminated in the reverse reaction. The rate constant k_{-3} [see Eq. (1)] may directly control the rate of intermediate exchange.

Table 1 shows the results of an experiment in which $[^{18}O]$-ATP labelled in the three terminal atoms of the γ-phosphoryl group was mixed with subfragment 1 and the extent of oxygen exchange in $M^{**}.ADP.P_i$ was compared at 37.6 ms and 2.35 s. Approximately one atom has exchanged

Table 1. Intermediate oxygen exchange in $M^{**}.ADP.P_i$ during transient phase of subfragment 1 ATPase

Time of acid quench	Calculated[a] a.p.e.	Observed a.p.e.	Corrected[b] observed a.p.e.	% ^{18}O-loss
37.6 ms	0.347	0.288	0.247	29
2.35 s	0.453	0.198	0.162	64

49 µM ATP (44.9 µM $[^{18}O]ATP + 4.1$ µM $[\gamma^{32}P]ATP$) was mixed in a quenched-flow apparatus with 43 µM subfragment 1[c] (reaction chamber concentrations) in 0.10 M KCl, 5 mM $MgCl_2$ and 50 mM tris adjusted to pH 8.0 with HCl at 22°C. Samples were quenched into 6% $HClO_4$ and carrier P_i. P_i was separated from ATP using the charcoal procedure and prepared for ^{18}O analysis as described previously (Bagshaw et al. 1975).

[a] a.p.e. (atom percent excess) calculated for dilution by carrier P_i if there was no exchange of the 3 terminal oxygen atoms.

[b] Corrections were made for $[^{18}O\ P_i]$ present initially in the $[^{18}O]$-ATP sample and for breakdown of ATP during the work up procedure.

[c] Nominally (from $\varepsilon_{280\ nm}$) subfragment 1 was 53 µM. In line with our normal experience its active site concentration was 43 µM. This concentration was based on the observation that 78.8% of the ATP was cleaved to $M^{**}.ADP.P_i$ at 196 ms on completion of the fast binding step. This means the subfragment 1 concentration which equals $[M^*.ATP]$ + $[M^{**}.ADP.P_i]$ at 196 ms was 87.5% that of the initial ATP concentrations when a value of 9 is taken for K_3. Subfragment 1 was prepared as described previously (Bagshaw and Trentham, 1973).

at 37.6 ms and two atoms at 2.35 s. It is not clear from these data whether the first two atoms exchange at the same rate and hence 37.6 ms is close to the half-time for this process or whether this exchange is biphasic and the first exchange is complete at 37.6 ms. To resolve this ambiguity the extent of the exchange must be followed through the millisecond time range. Such studies are in progress.

This result has certain implications. The time course of the exchange is such that there is scope for comparing the time course of exchange in the myosin and actomyosin ATPases and so test the Lymn-Taylor scheme [Eq. (2)]. The exchange kinetics are sufficiently slow to allow experiments to obtain more insight into the chemistry of the cleavage step. For example, it is likely that the hydrolysis of $M^*.ATP$ is the initial step in intermediate exchange [see Eq. (3)]. However this point can now be clarified by analysing the composition of the bound P_i within a few ms of mixing myosin and ATP in $H_2^{18}O$. It is important to do this not only because of its relevance to the hydrolysis mechanism but also because deductions about the values of k_{+3} and k_{-3} made from kinetic oxygen-isotope studies depend upon the validity of Eq. (3).

II. Reversibility of the Cleavage Step in the Actomyosin ATPase

The equilibrium constant of the cleavage step of the actin-subfragment 1 ATPase, K'_3, was measured in experiments that were similar to those used to measure K_3 in the subfragment 1 ATPase (Bagshaw and Trentham,

1973). In a typical experiment a solution of 1.55 mg/ml F-actin (37 μM in actin monomers) and 25 μM subfragment 1 was mixed with 12.5 μM (γ - ^{32}P) ATP and the reaction was quenched into 7% perchloric acid at 196 ms. [Based on a second order rate constant of 1×10^6 $M^{-1}s^{-1}$ for the actin subfragment 1 association with ATP (Lymn and Taylor, 1971), the binding reaction will be 95% complete at 196 ms]. Comparison of the amount of [^{32}P]P$_i$ and [γ-^{32}P] ATP in the reaction products then, in principle, gives K'$_3$ when allowance is made for breakdown of the subfragment 1 - products complex (controlled by the rate constant k'$_{+4}$ of Eq. (2). Table 2 records that 25.7% of the initial ATP was present at 196 ms. It is important to establish to what extent this ATP was bound at the active site as opposed to being free in the medium or bound at other sites on the proteins. This was done in an experiment which is similar to the 'cold chase' experiment carried out to establish the nature of protein bound ATP during subfragment 1 ATPase activity (Bagshaw and Trentham, 1973). Instead of quenching the reaction mixture into perchloric acid, the solution was added to a relatively concen-trated solution of non-radioactive ATP at 196 ms. [γ-^{32}P] ATP not already bound at the active site is thereby prevented from binding and being hydrolysed. [γ-^{32}P] ATP present at the active site is potentially cap-able of being hydrolysed although it may dissociate from the active site as [γ-^{32}P] ATP. The presence of non-radioactive ATP will then pre-vent its hydrolysis. Table 2 gives the results of this experiment and shows that 10.9% of the [γ-^{32}P] ATP was either not bound or dissociated from the active site during the 'cold chase'. This means that at least (25.7 - 10.9) = 14.8% of the [γ-^{32}P] ATP was bound to the active site when the reaction was quenched at 196 ms into perchloric acid. When a similar set of experiments was carried out with the same subfragment 1 preparation but in the absence of actin almost identical results were obtained (Table 3).

The catalytic centre activity of the Mg^{2+}-dependent actin-subfragment 1 ATPase, measured in a pH stat at 20°C and pH 8 in the presence of 1.55 mg/ml F-actin in 2 mM MgATP, 75 mM KCl and 1 mM tris, was 0.35 s^{-1}.

Table 2. Protein-bound ATP during the actin-activated subfragment 1 ATPase

Quench time	Chasing time	% labelled ATP
0		100
196 ms		24.0, 27.4 Mean: 25.7
60 s	196 ms	11.1, 10.7 Mean: 10.9
60 s		2.8

A solution of 1.55 mg/ml F-actin (prepared according to Lehrer et al., 1972, nucleotide content see Table 4) and 25 μM subfragment 1 (reac-tion chamber concentrations) was mixed with 12.5 μM [γ-^{32}P]ATP in 50 mM KCl, 2 mM MgCl$_2$ and 50 mM tris adjusted to pH 8.0 with HCl at 21°C. For the 196 ms quenching the reagents were mixed in a quenched flow apparatus and ejected into an equal volume of 7% HClO$_4$ at 196 ms. The products were analysed as described previously. In the cold chase experiments the reagents were ejected into ATP (final concentration 5 mM) at 196 ms and the reaction was quenched with HClO$_4$ at 60 s. The zero and 60 s time points were obrained as described previously. Each value has been corrected for decomposition of ATP during the quenching procedure which was 5% of the ATP present at the time of quenching.

Table 3. Protein-bound ATP during the subfragment 1 ATPase

Quench time	Chasing time	% labelled ATP
0		100
2.0 s		24.0, 24.0 Mean: 24.0
60 s	2.0 s	8.9, 9.1 Mean: 9.0
60 s		2.1

The experiment described in Table 2 was repeated using the same sub-fragment 1 preparation but with no actin present. The 2 s quenching and 'cold chase' measurements were made as described previously.

The catalytic centre activity increased linearly with actin concentration in the range 0-3 mg/ml F-actin - a result consistent with other steady-state kinetic studies (Moos, 1972).

Marston and Weber (1975) have estimated the actin-subfragment 1 dissociation constant as about 10^{-8} M. This means the concentration of free subfragment 1 in the experiment of Table 2 was less than 10^{-7} M. This conclusion was born out using turbidity change as a probe for the extent of actin-subfragment 1 complex formation, adding aliquots of subfragment 1 to 0.11 mg/ml F-actin at pH 8 and observing the break point in the turbidity vs. [subfragment 1] graph. This analytical procedure also served as a check on the relative concentrations of actin to subfragment 1 used in this series of experiments (Tables 2 and 4). The turbidimetric analyses were performed in a Farrand Mark 1 spectrofluorimeter. Light at 350 nm was scattered by the turbid solution and measured at 90° to the incident beam. This technique suggested by Dr. A.G. Weeds is more sensitive than measurement of transmission changes conventionally used to monitor actomyosin complex formation. Hence the protein bound [γ-^{32}P] ATP did not arise because of free subfragment 1 in the medium at zero reaction time.

As mentioned above, ATP bound at the active site of subfragment 1 during ATPase activity exchanges the oxygen atoms of its γ-phosphoryl group with the solvent while ATP in the medium does not. This oxygen exchange capacity can be used as an indicator of whether the ATP isolated in the actin-subfragment 1 experiment was bound at the active site. Table 4 gives the results of two experiments in which the oxygen was examined. 61% oxygen exchange was found to have occurred in the 3 oxygen atoms of the γ-phosphate groups of the isolated ATP indicating that a significant fraction of this ATP was bound at the active site of the actin-subfragment 1 ATPase. Rather more oxygen exchange (83%) was found in the isolated P_i. This difference in the extent of exchange indicates that some of the isolated ATP was not protein-bound at the time of quenching the reaction. This is consistent with the results of Table 2. 75% oxygen exchange was detected in the 3 oxygen atoms of the γ-phosphate group of protein bound ATP isolated in comparable experiments with subfragment 1 alone (Bagshaw et al., 1975). The extent of oxygen exchange in the ATP isolated from the subfragment 1 and actin-subfragment 1 experiments is the same within experimental error (see Bagshaw et al., 1975 for a discussion of the errors).

Throughout this series of experiments K_3 and K'_3 were up to a factor of 2 smaller than the value of K_3 of 9 found by Bagshaw and Trentham (1973). This probably reflects the large scale and perhaps therefore less pure protein preparations needed in this series of experiments

Table 4. Oxygen exchange in the γ-phosphoryl group of ATP following its reaction with a molar excess of actin-subfragment 1

Experiment number	Ratio $[P_i]/[ATP]$ present at 196 ms[a]	Calculated a.p.e.[b]	Observed a.p.e.	% ^{18}O loss
1	6.0	0.269	0.101	62
2	5.3	0.275	0.109	60

In Experiment 1 1.55 mg/ml F-actin[c] and 30 μM subfragment 1 (reaction chamber concentrations) were mixed in a quenched-flow apparatus for 196 ms with 15.3 μM ATP (150 μM $[^{18}O]$ATP labelled in 3 terminal oxygen atoms and 0.3 μM $[γ-^{32}P]$ATP; total 1.00 μmole ATP) in the same solvent as in Table 2. The reaction mixture was ejected into 7% perchloric acid containing carrier P_i, adjusted to pH 4 and carrier ATP added. The ATP and P_i in the mixture were separated by the charcoal procedure and analysed for ^{18}O content as described previously (Bagshaw et al., 1975). The same solvent and similar reactant concentrations were used in experiment 2. The % ^{18}O losses in the isolated P_i were 82% and 84% in experiments 1 and 2 respectively.

[a] The ratio $[P_i]/[ATP]$ present at 196 ms was calculated from the measurement of ^{32}P following the separation of the ATP and P_i on charcoal.

[b] Calculated a.p.e. for dilution by carrier ATP if there was no exchange of the 3 terminal oxygen atoms.

[c] ATP in the supernatant following polymerisation from G-actin was removed by pelleting and resuspending the F-actin. ADP was shown to be present in a 1.0:1 molar ratio with the F-actin monomer. This ADP was released into the medium in the acid quench and so was taken into account in evaluation of calculated a.p.e.

to isolate sufficient protein-bound ATP from the quenched reaction for mass spectroscopy. The relatively large amount of ATP detected in the 'cold chase' experiment of Table 3 was noted occasionally by Bagshaw and Trentham (1973).

These results (Tables 2, 3 and 4) provide support for a principal tenet of the Lymn-Taylor hypothesis that the step M*.ATP.$\rightleftharpoons$M**.ADP.P_i is common to the Mg^{2+}-dependent myosin and actomyosin ATPase mechanisms. Nevertheless it is desirable to carry out these actin-subfragment 1 experiments at lower ionic strength so that the reaction conditions are those appropriate to maximum actomyosin ATPase activity. These experiments will be more difficult technically as the lifetime of protein-bound ATP will be shorter and hence a greater proportion of isolated ATP is likely to be ATP from the solvent. However in principle such experiments are feasible now that the basic methodology has been established.

III. The Binding of ATP to Arterial Actomyosin

The kinetics of ATP binding to actomyosin have generally been monitored by measuring the rate of the accompanying turbidity change (Finlayson et al., 1969). In contrast to the kinetics observed when ATP reacts with subfragment 1 [Eq. (1)], the kinetics of ATP binding to rabbit skeletal actomyosin at 20°C and pH 8 in the presence of $MgCl_2$ show a first order dependence on ATP concentration up to observed rates in excess of 700 s^{-1}, the maximum rate measurable using absorption stopped flow equipment (Lymn and Taylor, 1971). Nevertheless the observed second order rate constant for the process, 1.6×10^5 $M^{-1}s^{-1}$ in 0.5 M KCl, is sufficiently slow to suggest that this may represent a complex function of rate constants possibly related to a two-step binding process as in Eq. (1).

The rates of elementary processes associated with the Mg^{2+}-dependent myosin and actomyosin ATPases of smooth muscle are likely to be much slower than the rates of the corresponding processes of the ATPases isolated from fast muscle. This is indicated by the relatively low specific activities of the ATPases, (Russell, 1973) transient kinetic studies of the arterial myosin ATPase (Mrwa and Trentham, 1975) and the physiological properties of smooth muscle. If the ATPase mechanisms of skeletal and smooth muscles are broadly similar except for the rate constants, there is more chance of detecting a two-step binding process, if present, with arterial actomyosin.

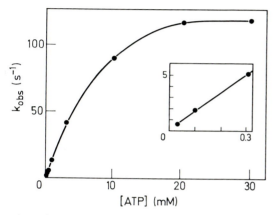

Fig. 1. Dependence of the rate of arterial actomyosin dissociation on ATP concentration (the inset covers the low range of ATP concentration). Various concentrations of MgATP were mixed with 1.65 mg/ml arterial actomyosin prepared from pig carotids (reaction chamber concentrations) in 0.5 M KCl, 5 mM Mg Cl$_2$ and 50 mM tris adjusted to pH 8.0 with HCl (pH adjustment at 20°C). A stock MgATP solution was prepared by mixing disodium ATP with 1.5 fold molar excess of MgCl$_2$ and adjusting to pH 8 with KOH. The reaction was carried out in an absorption stopped flow apparatus at 5°C and the decrease in absorption accompanying the decrease in turbidity was observed at 350 nm as an exponential process whose rate was k_{obs}.

Control experiments were carried out mixing 20 μM ATP and actomyosin at various KCl and MgCl$_2$ concentrations to measure the maximum influence of ionic strength variation on the observed second order reaction rate. The maximum effect of compensating for the change of ionic strength would be to increase the rate at 30 mM MgATP by a factor of 1.4

Fig. 1 shows the kinetics of ATP binding to arterial actomyosin. The observed rate reaches a plateau of 120 s^{-1}. The experiment was repeated with five different preparations of arterial actomyosin and in each case a plateau was observed. Care was taken to show that the plateau was not a consequence of changing ionic strength at the different ATP concentrations as decribed in the figure legend.

Two kinetic schemes, compatible with these results, will be considered:

$$AM + ATP \underset{k'_{-1}}{\overset{k'_{+1}}{\rightleftharpoons}} AM.ATP \underset{k'_{-2}}{\overset{k'_{+2}}{\rightleftharpoons}} M.ATP + A \qquad (4)$$

$$AM + ATP \underset{k''_{-1}}{\overset{k''_{+1}}{\rightleftharpoons}} AM.ATP \underset{k''_{-2}}{\overset{k''_{+2}}{\rightleftharpoons}} AM^*.ATP \underset{k''_{-3}}{\overset{k''_{+3}}{\rightleftharpoons}} M^*.ATP + A \qquad (5)$$

In Eq. (4) either (a) $k'_{+2} = 120$ s^{-1} and $\frac{k'_{+1}k'_{+2}}{k'_{-1}} = 4 \times 10^4$ M^{-1}s^{-1}
or (b) $k'_{+1} = 4 \times 10^4$ M^{-1}s^{-1} and $k'_{+2} = 120$ s^{-1}. In Eq. (5) $k''_{+2} = 120$ s^{-1} and $\frac{k''_{+1}k''_{+2}}{k''_{-1}} = 4 \times 10^4$ M^{-1}s^{-1}. There are a number of reasons
indicating that Eq. (5) is preferable as a plausible mechanism though none of these is compelling. (1) Low concentrations of ATP are effective in causing actomyosin dissociation and the data of Table 2 indicate the cleavage step of the actomyosin ATPase is readily reversible. It follows that the overall equilibrium constant of Eqs. (4) and (5) is markedly in favour of A + M.ATP (or M*.ATP) formation. The experiment which follows using adenosine 5'-(3-thiotriphosphate), ATP(γS), supports this conclusion. So either k'_{-2} is an abnormally small bimolecular association rate constant for the two proteins [Eq. (4) alternative (a)] or k'_{+1} is an abnormally small bimolecular association rate constant for a ligand protein interaction [Eq. (4) alternative (b)]. Furthermore if alternative (b) were correct, there would be a detectable lag phase in the turbidity change. No lag phase was observed. (2) Comparisons between cardiac and skeletal myosins suggest that it is the kinetics of isomerizations which distinguish muscle types (Bagshaw et al., 1974). According to Eq. (5) k''_{+2} would be the rate constant characteristic of the muscle type. (3) If the process M*.ATP$\rightleftharpoons$M**.ADP.P$_i$ is common to both myosin and actomyosin mechanisms, then myosin probably dissociates from actin as M*.ATP. (However we cannot rule out that actin maintains myosin as an isomer analogous to M*.ATP in the actomyosin complex; i.e. as AM* rather than AM.)

The kinetics of the MgATP(γS)-induced dissociation of arterial actomyosin were followed in the stopped-flow apparatus at 5°C. The second order rate constant for the process was 1.2×10^3 M^{-1}s^{-1} in the same solvent as in Fig. 1. Provided [ATP(γS)]>[actomyosin] the amplitude of the absorption change was constant so that, as with ATP, the equilibrium constant for the process markedly favoured arterial actomyosin dissociation. Since ATP(γS) is only cleaved slowly by rabbit skeletal myosin (Bagshaw et al., 1972) it is likely that nucleotide binding alone is effective in promoting arterial actomyosin dissociation.

D. Conclusion

The above results support Eq. (2) insofar as it describes intermediates leading to the formation of M**.ADP.P$_i$ in the actin-activated Mg^{2+}-dependent subfragment 1 ATPase. It is likely that an additional intermediate AM*.ATP should be included but the evidence for this is not yet compelling.

Acknowledgments. We are grateful to Professor H. Gutfreund and Professor P.D. Boyer for helpful discussions. ^{18}O-analyses were carried out in Professor Boyer's laboratory. We thank the National Science Founda-

51

tion, the Atomic Energy Commission, U.S.A., the Science Research
Council, U.K., the European Molecular Biology Organisation and the
Deutsche Forschungsgemeinschaft (SFB 90) for financial support.

References

Bagshaw, C.R., Eccleston, J.F., Eckstein, F., Goody, R.S., Gutfreund,
  H., Trentham, D.R.: Biochem. J. 141, 351-364 (1974).
Bagshaw, C.R., Eccleston, J.F., Trentham, D.R., Yates, D.W., Goody,
  R.S.: Cold Spring Harbor Symp. Quant. Biol. 37, 127-135 (1972).
Bagshaw, C.R., Trentham, D.R.: Biochem. J. 133, 323-328 (1973).
Bagshaw, C.R., Trentham, D.R., Wolcott, R.G., Boyer, P.D.: Proc. Nat.
  Acad. Sci. 72, 2592-2596 (1975).
Finlayson, B., Lymn, R.W., Taylor, E.W.: Biochemistry 8, 811-819
  (1969).
Lehrer, S.S., Nagy, B., Gergely, J.: Arch. Biochem. Biophys. 150,
  164-174 (1972).
Levy, H.M., Koshland, D.E.: J. Biol. Chem. 234, 1102-1107 (1959).
Lowey, S., Slayter, H.S., Weeds, A.G., Baker, H.: J. Mol. Biol. 42,
  1-29 (1969).
Lymn, R.W., Taylor, E.W.: Biochemistry 9, 2975-2983 (1970).
Lymn, R.W., Taylor, E.W.: Biochemistry 10, 4617-4624 (1971).
Mannherz, H.G., Schenck, H., Goody, R.S.: Eur. J. Biochem. 48, 287-295
  (1974).
Marston, S., Weber, A.: Biochemistry 14, 3868-3873 (1975).
Moos, C.: Cold Spring Harbor Symp. Quant. Biol. 37, 137-143 (1972).
Mrwa, U., Trentham, D.R.: Hoppe Seyler's Z. Physiol. Chem. 356, 255
  (1975).
Russell, W.E.: Eur. J. Biochem. 33, 459-466 (1973).
Sartorelli, L., Fromm, H.J., Benson, R.W., Boyer, P.D.: Biochemistry
  5, 2877-2884 (1966).
Weber, H.H., Portzehl, H.: Prog. Biophys. Biophys. Chem. 4, 60-111
  (1954).
Wolcott, R.G., Boyer, P.D.: Biochem. Biophys. Res. Comm. 57, 709-716
  (1974).

Discussion

Dr. Hess: Is the mitochondrial ATPase similar to the actomyosin ATPase?

Dr. Trentham: The mitochondrial ATPase is being investigated very actively by Boyer's group in California. A principal finding (Boyer et
al., 1973) is that, as with myosin, it is possible to characterize
bound ATP in submitochondrial particles even in the presence of uncouplers such as 2,4-dinitrophenol. In addition these workers observed
that submitochondrial particles catalyse oxygen exchange between P_i
and water in the presence of uncouplers. This exchange is consistent
with a mechanism which includes a readily reversible ATP cleavage step
similar to that identified in the myosin ATPase (Cross and Boyer, 1975).

Dr. Gergely: Could you comment on the so-called "refractory state"
postulated by Eisenberg and his colleagues?

Dr. Trentham: It appears to be an intermediate of the actomyosin ATPase,
probably a myosin product complex of some sort, which cannot bind actin
or binds it very weakly. I think it is an area of research to try and
find out if we can detect this intermediate by methods other than they
have used.

Dr. Chaplain: Couldn't it be that there exists a magnesium-ADP-ATP-complex of the two myosin heads which represents that refractory state? Such a double complex would also account for the fact that at high ATP concentrations a decrease in the rate constant of ADP release can be observed (Chaplain and Gergs, FEBS Letters $\underline{43}$, 277-280, 1974). Did you have evidence for a negative cooperativity between the two heads?

Dr. Trentham: The whole problem of the interaction of the two heads is another major problem of work, but I think that heavy meromyosin subfragment 1 itself exhibits the refractory state (Eisenberg and Kielley, 1972).

Dr. Helmreich: I would like to make a brief comment with regard to your two-step model where one step is an isomerization. Have you tried to verify this mechanism more directly by using a chromophoric non-utilizable ATP-analogue, for example an (etheno)ATP(γS)?

Dr. Trentham: It is possible to take chromophoric analogues of ATP and see if the observed association rate constant reaches a plateau at high analogue concentrations. We have tried to do this with thioATP (6-mercapto-9-β-ribofuranosylpurine 5'-triphosphate) which quenches protein fluorescence on binding to myosin. The rate of the observed exponential fluorescence change appears to reach a plateau whose value equals that observed when ATP is substrate. The initially formed binary complex between thioATP and the protein has a smaller association constant than that characteristic of ATP. Therefore higher concentrations of thioATP are required to observe the plateau. At these high nucleotide concentrations significant inner filter effects are present resulting in a relatively low signal to noise ratio in our experiments.

Dr. Wallenfels: The rapid and extensive ^{18}O-exchange on the level of the M-ADP-P_i-complex points to a different chemical structure of this important intermediate of muscle contraction. If water enters into the M-ATP-complex, it may form a fifth ligand of the γ-phosphorous atom building up a complex of trigonal bipyramidal structure with phosphate in the centre of it. The oxygen atoms then may be randomized by pseudo-rotation.

Dr. Trentham: The idea of pseudorotation in the myosin ATPase mechanism has been developed by Young et al. (1974). We need more experimental evidence to distinguish the various mechanisms of oxygen exchange that have been proposed.

Boyer, P.D., Cross, R.L., Momsen, W.: Proc. Nat. Acad. Sci. $\underline{70}$, 2837-2839 (1973).
Cross, R.L., Boyer, P.D.: Biochemistry $\underline{14}$, 392-398 (1975).
Eisenberg, E., Kielley, W.W.: Cold Spring Harbor Symp. Quant. Biol. $\underline{37}$, 145-152 (1972).
Young, J.H., McLick, J., Korman, E.F.: Nature $\underline{249}$, 474-476 (1974).

The Kinetics of Cross-bridge Turnover

R. J. Podolsky

A. Structure of the Sarcomere

I would like to describe some methods that have been used to study the kinetics of cross-bridge turnover in muscle cells, and to tell you about some of the results that have been obtained. The structural basis of the problem is shown in Fig. 1. This electron micrograph of the

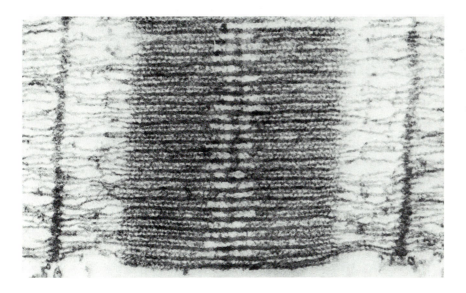

Fig. 1. Electron micrograph of striated muscle fibre. (From Huxley, 1957)

sarcomere, which was published by H. Huxley (1957) nearly twenty years ago, shows the major components of the contractile mechanism. These are the thick, myosin-containing filaments; the thin, actin-containing filaments; the Z line, which connects the thin filament arrays in each myofibril; and the connections between the thick and thin filaments, the cross-bridges. The elementary contractile unit is the half sarcomere, which is about 10,000 Å in length. The length of a cross-bridge is close to 100 Å, or about 1% of the length of the half sarcomere.

Since a muscle fibre can shorten considerably more than 1% of its length, the cross-bridges must turn over many times during a contraction. Information about the cross-bridge "reach," the rate functions for making and breaking the bridge, and the degree of interaction be-

tween adjacent bridges can be derived from an analysis of the motion
of the fibre under various conditions.

B. Physiological Preparations

Fig. 2 shows preparations that have been used in these studies. The
earliest experiments were done with whole muscles (top panel). A force

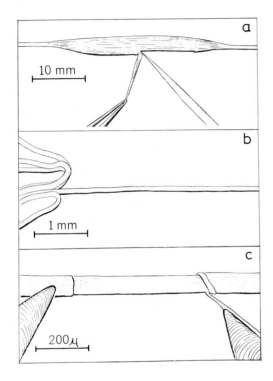

Fig. 2a-c. Muscle fibre preparations. (a) Intact muscle. Panel
also shows a step in the removal
of a bundle of muscle fibres. (b)
(b) An isolated, single fibre.
(c) Preparation of a skinned
muscle fibre. (From Podolsky,
1968)

transducer was attached to one end and a displacement transducer to
the other. The advantage of this preparation is that it generates
many grams of force; the disadvantage is that the fibres in a muscle
are not all the same length, so that each fibre shortens somewhat dif-
ferently as the muscle shortens. This difficulty can be overcome by
working with either a small bundle of fibres or a single fibre (centre
panel). For most physiological experiments, preparations a and b are
activated by electrical stimulation, which releases calcium from the
sarcoplasmic reticulum into the fibre volume.

A simple preparation in which the chemical conditions of contraction
can be controlled experimentally is the skinned muscle fibre (Natori,
1954). This is made by removing the cell membrane from a single fibre
by microdissection (lower panel). The myofibrils can then be bathed
in a solution chosen by the experimenter. The preparation is activated
by raising the calcium ion concentration from pCa 8 to about pCa 5
(Hellam and Podolsky, 1969).

C. Independent Force Generators

It is generally supposed that each cross-bridge acts as an independent
force generator. Several lines of evidence suggest that this is the
case. A. Huxley and his collaborators (Gordon et al., 1966) made a
careful study of the relation between sarcomere length and contractile
force, and they found that force was proportional to the degree of
filament overlap. This result is clearly consistent with the idea that
the cross-bridges act independently. However, it leaves open the possibility that the force produced by a given cross-bridge is affected
by the state of an adjacent actin site. Information on this point was
obtained from studies by Teichholz and myself with skinned muscle
fibres (Podolsky and Teichholz, 1970), where the degree of activation,
and presumably the distance between activated actin sites, was varied
while the filament overlap remained constant.

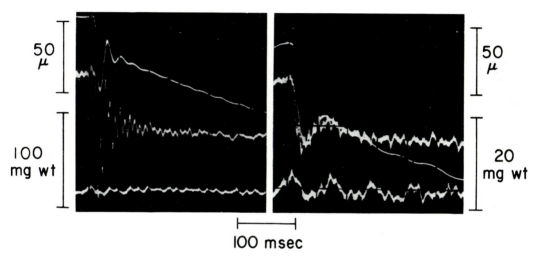

Fig. 3a and b. Contraction of a skinned frog muscle fibre segment at
different calcium levels. Upper trace, displacement; lower trace,
force; segment length, 1.0 mm. (a) pCa 5. (b) pCa 6.75. (From Podolsky
and Teichholz, 1970)

The critical experiment is shown in Fig. 3. The lower trace is force
and the upper trace is displacement. In the panel on the left, the
fibre was activated at pCa 5. After the isometric force became steady,
the load was reduced to about half the isometric value, and the fibre
shortened at a steady rate. In the right-hand panel, the fibre was
activated at pCa 6.75 and the isometric force dropped about 5-fold.
However, when the load was reduced and the fibre was allowed to shorten,
the steady contraction velocity was the same as at pCa 5. This experiment (which was repeated recently at high time resolution; see below)
suggests that cooperative effects along the length of a myofilament
do not affect the kinetic parameters of cross-bridge turnover, and it
provides additional evidence that the cross-bridges act independently.

D. Contraction Transients

I would like to turn now to some results obtained from the contraction
transients (Podolsky, 1960). Fig. 4 shows a set of experiments carried
out with a frog sartorius muscle 33 mm in length. The lower trace in
each panel is the force. The muscle was activated by electrical stimulation and after force reached the steady isometric level, the load
was quickly reduced to a certain fraction of this value. The top trace
in each panel gives the motion of the muscle. The fibres shorten very
quickly as force changes; this very fast initial motion is called the

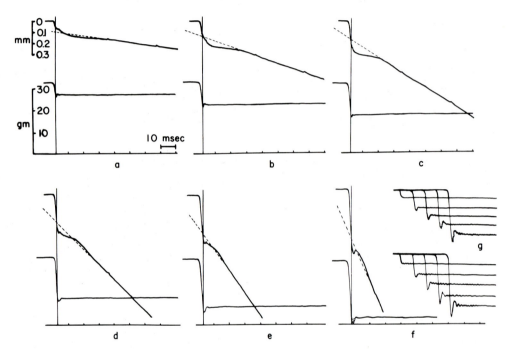

Fig. 4a-g. Isotonic velocity transients of intact muscle fibres following step changes in load. Upper trace, displacement; lower trace,
force. Relative force step: (a) 0.16, (b) 0.31, (c) 0.45, (d) 0.60,
(e) 0,74, (f) 0.90; frog sartorius muscle at 0°C, (g) control experiment with spring. (From Podolsky, 1960)

quick displacement. The quick displacement that reduces the force to
less than half of its initial value is close to 0.3 mm, or about 1%
of the muscle length. Some of this displacement is taken up by the
ends of the preparation, but most of it is distributed along the
length of the fibre (Jewell and Wilkie, 1958). At the level of the
half sarcomere, this corresponds to a displacement of the order of
100 Å, which is very close to the length of the cross-bridges in
electron micrographs. This result suggests that the reach of a cross-
bridge is the same order as its length, and that the quick displacement associated with a change in force is to a large extent due to
release of stress in the cross-bridges. However, it does not rule out
the possibility that some of the quick displacement is due to a change
in strain of a passive structure within the sarcomere, the Z line, for
example.

After the initial quick displacement, isotonic shortening does not become steady for at least 10 msec. If the duration of the presteady motion is a measure of the cross-bridge turnover time, which is a reasonable assumption, this response indicates that the average turnover time of a cross-bridge during shortening is of the order of 10 msec. Panel f, where the load is small and the fibre shortens at close to the maximum steady value, shows that the fibre shortens about 1% of its length, or 100 Å/half sarcomere, in 10 msec. In other words, the contraction appears to be caused by displacements of the order of 100 Å occuring in times of the order of 10 msec. Thus the time course of the motion after a force step also supports the idea that the molecular scale of cross-bridge movement is of the order of 100 Å.
If we examine the isotonic transients more closely, we see that the motion resembles a damped sine wave superimposed on steady shortening. The initial velocity after the force step is considerably faster than the steady motion. Several explanations have been put forward to account for these early rapid motions, but before discussing these I would like to explain how the contraction transients can be dealt with analytically.

E. Huxley Model (1957)

The first quantitative method for analyzing the motion of muscle fibres in terms of cross-bridge parameters was put forward by A. Huxley in 1957. Huxley divided the population of actin sites in a sarcomere into two states: those that were part of a cross-bridge, and those that were not. The partition of sites between these two states was described by a distribution function n(x), where x is the position of an actin site relative to the nearest projection on the myosin filament. New cross-bridges were made from unattached sites at a rate f(x), and existing cross-bridges broke at a rate g(x). Cross-bridges at a given x developed force, k(x); this does not require that force by present at the moment the cross-bridge is made, but it does imply that the force generating process is very rapid on the physiological time scale.

Huxley showed that the correct force-velocity relation could be obtained by using a moderate value for f. He also accounted for the energetics of contraction by assuming that an ATP molecule was hydrolysed each time a cross-bridge turned over.

Fig. 5 shows the motion computed from the Huxley model following changed in load (Civan and Podolsky, 1966). The duration of the contraction transients is of the order of 10 msec, as is found experimentally. However, the early high velocity motion is absent. When we noticed this, we wondered whether the correct motion could be obtained from the Huxley model simply by changing the rate functions for cross-bridge turnover.

According to the model, a quick displacement relieves the stress in the existing cross-bridges and brings additional actin sites into positions where it is possible for them to make new cross-bridges. The velocity of the early isotonic motion depends on the rate at which these new bridges are made. We found that we could simulate the early rapid motion very accurately by supposing that the rate constant for making cross-bridges is an order of magnitude greater than that used by Huxley (Podolsky and Nolan, 1973). If this were indeed the case, we would expect the number of cross-bridges in the fibre to increase significantly in the first few milliseconds after the quick displacement.

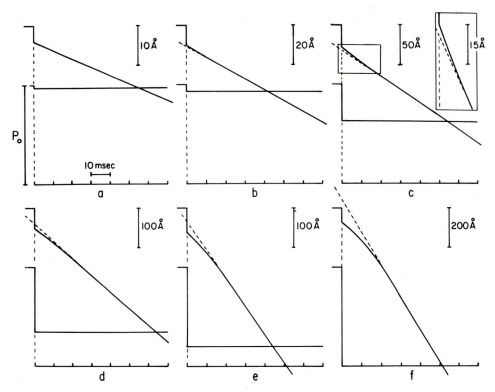

Fig. 5a-f. Response of A.F. Huxley (1957) contraction model to step changes in load. (From Civan and Podolsky, 1966)

F. Stiffness Measurements

One way of assaying the number of cross-bridges in a fibre is to measure its stiffness. The principle of the method is illustrated in Fig. 6. Suppose a spring with force constant k is strained an amount x. Then, as shown on the left, the force P will be kx. The stiffness of the <u>system</u> is found by letting it shorten an amount $\Delta \underline{x}$, measuring the new force P', and taking the ratio $(P'-P)/\Delta \underline{x}$.

If the system consists of two springs in parallel, each with force constant k, it will be twice as stiff. The right-hand side of Fig. 6 shows the steps in the stiffness measurement. Force P is exerted when both springs are strained an amount $\frac{x}{2}$. When the system is displaced by $\Delta \underline{x}$, the new force P' is $k\underline{x}-2k\Delta \underline{x}$. The stiffness in this case turns out to be 2k.

Therefore, for the case of linear springs, the number of springs between two filaments is assayed by the stiffness of the overall system.

In the case of muscle cells, the situation is more complex since the cross-bridges are continually making and breaking. The stiffness measurement must be made more quickly than the cross-bridge distribution is changed by turnover. Consider two extreme cases. If the stiffness

$$P \qquad kx \qquad k\frac{x}{2} + k\frac{x}{2} = kx$$

$$P' \qquad k(x - \Delta x) \qquad k\left(\frac{x}{2} - \Delta x\right) + k\left(\frac{x}{2} - \Delta x\right) = kx - 2k\Delta x$$

$$\frac{P' - P}{\Delta x} \qquad k \qquad\qquad 2k$$

Fig. 6. Principle of stiffness measurement. For linear springs, the stiffness is proportional to the number of springs that act in parallel

measurement were made slowly relative to the turnover rate, the muscle force would follow the length-tension curve and the fibre would appear to be considerably more compliant than during a quick displacement. On the other hand, if the stiffness measurement were made very quickly, to eliminate the influence of cross-bridge turnover, two other phenomena must be considered. One is the finite transmission time of mechanical impulses down the length of a muscle fibre, which sets a natural limit to the speed with which stiffness measurements can be made by the quick displacement method. Schoenberg et al. (1974) recently measured the mechanical propagation velocity for activated frog muscle fibres, and found that a signal travelled down the length of a 10 mm fibre in about 60 μsec. To eliminate the influence of propagation effects on stiffness measurements the displacement step must be spread out over many transmission times; for a 10 mm fibre, a stepping time of at least several tenths of a millisecond is required.

Viscous forces may also play a role in rapid stiffness measurements. These forces tend to couple the two sets of myofilaments to each other, thereby making the fibre appear stiffer than it actually is.

A final possibility that should be kept in mind is that cross-bridges may break during a displacement step, due to the change in strain. This seems to happen to a certain extent during rapid stretches (Sugi, 1972).

The overall effect of these factors is to introduce a velocity dependence in the stiffness measurement. Is this an important effect quantitatively? One way of finding out experimentally is to see how stiffness depends on the duration of the test step. If it were possible to find a "window" in which stiffness were independent of stepping time, these factors could be discounted. To date, however, this has not been done: the apparent stiffness of the sarcomere has tended to increase as the technology for applying test steps has improved. This is a complicated problem which is still being sorted out.

G. Contraction Distance to the Null Point

Another feature of the isotonic motion that influenced out thinking is illustrated in Fig. 7. These are force step experiments made with a small bundle of frog fibres (Civan and Podolsky, 1966). We noticed that the amount of shortening between the end of the force step and

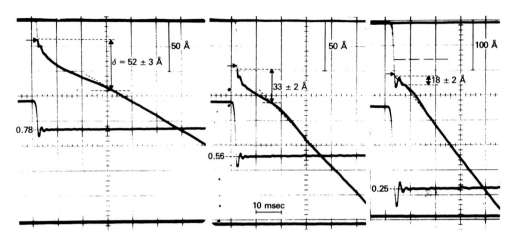

Fig. 7. Contraction distance to the null point. Bundle of frog fibres at 3°C. Records from Fig. 3 of Civan and Podolsky (1966); displacement scale, Å per half sarcomere. The fibre length at the start of isotonic contraction was estimated by two methods. In one, the results of which are marked by the horizontal arrows alongside the displacement traces, the displacement overshoot was calculated from the product of the chord compliance (Appendix to Civan and Podolsky, 1966) and the amplitude of the initial force overshoot. In the second, the results of which are indicated by the horizontal dotted lines, the displacement over-shoot was calculated from the dynamic compliance (Appendix to Civan and Podolsky, 1966). Both methods gave similar results

the null point (the point at which the actual motion first intersects the back extrapolation of the steady motion) decreased as the force step increased (Podolsky et al., 1969). The upper horizontal dotted line in each panel is an estimate of the length of the fibre at the end of the quick displacement, and the line below this marks the level of the null point. The distance between these two points is measured by the double-headed arrow. When the relative load was 0.78 (left-hand panel) the contraction distance to the null point was 52 Å/half sarco-mere. For a load of 0.56, the contraction distance decreased to 33 Å/ half sarcomere (centre panel). In the right-hand panel the relative load was lowered further to 0.25; note that the gain for this dis-placement trace is reduced two-fold. The contraction distance to the null point was estimated to be 18 Å/half sarcomere or nearly half of that in the centre panel. The interrupted line in this panel is where the isotonic motion would originate if the contraction distance to the null point were the same as it is for a load of 0.75; this is clearly beyond the range of experimental error.

What does this mean? The contraction velocity just before the null point is considerably less than the steady value; the null point seems to mark the onset of a process that accelerates the motion. The fact that the isotonic contraction distance to this point decreased as the force step increased suggested to us that the rate function for break-ing cross-bridges has the spatial distribution shown by the g-function in Fig. 8 (Podolsky and Nolan, 1973).

The upper part of the figure is for orientation. A site on the actin filament is followed as it moves towards the left past a projection on the myosin filament. The origin of the coordinate system is chosen to be the position of the actin site where the cross-bridge force,

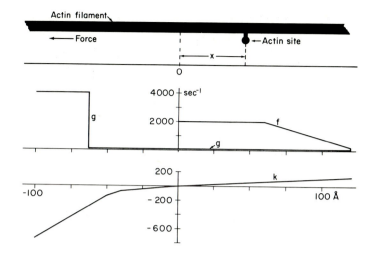

Fig. 8. Rate functions and force function for cross-bridge turnover according to Podolsky and Nolan (1973)

shown on the lower part of the slide, is zero. f is the rate function for making cross-bridges; the cross-bridge distribution function during isometric contraction has the same spatial distribution as this function. A quick displacement of the myofilaments will shift the cross-bridge distribution to the left, and the distribution function will continue to move in that direction when the cross-bridges turn over. As bridges generating negative force accumulate, the motion will slow down. However, when the left edge of the distribution function enters the region where g is very large, the negative force bridges will break quickly and the motion will accelerate. The amount of iso-tonic contraction that takes place before this acceleration occurs will depend on how far the initial distribution has been shifted by the quick displacement: the greater the initial shift, the smaller the required amount of isotonic contraction. This is consistent with the behavior of the living muscle seen in Fig. 7.

H. The Huxley-Simmons Model

An alternative explanation for the rapid early phase of motion has recently been put forward by Huxley and Simmons (1971). These authors suggested that the early motion after a quick displacement is due to the reorientation of existing cross-bridges rather than the formation of new cross-bridges.

This idea is diagrammed on the right of Fig. 9 (Huxley, 1974). The essential features are that (1) force is generated by the step-wise rotation of the myosin head relative to the thin filament; (2) this rotation strains a compliant element connecting the head to the back-bone of the thick filament, and (3) the time scale of this movement is about 1-2 msec. This contrasts with an earlier proposal of H. Huxley (1969) shown on the left, whereby force is generated by the tendency of the myosin head to rotate relative to the thin filament. According to Huxley and Simmons, a quick displacement of the myofilaments causes S_2 to shorten, while according to H.Huxley, quick displacement would cause the myosin head to rotate. It is important to realize that force could be developed according to the mechanism on the right, but if the time constant for S_1 rotation were much faster than a millisecond,

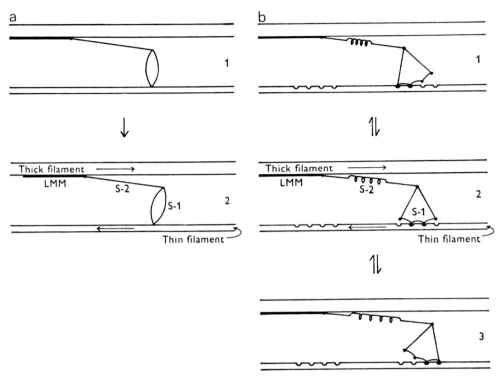

Fig. 9a and b. Hyopthetical force generating mechanisms. Myosin molecules have three components, light meromyosin (LMM) and subfragments
1 and 2 (S_1 and S_2). The body of the thick filament is made up of
LMM; the thick filament projections consist of S_1 and S_2, which are
thought to be flexibly connected. Cross-bridges are formed by the interaction of S_1 with the thin filament. Force can be generated by
either (a) the tendency of S_1 to rotate relative to the thin filament,
which stresses (but does not significantly stretch) S_2 (H.E. Huxley,
1969) or (b) the stepwise reaction of S_1 with a series of closely
spaced thin filament sites, which causes S_1 to rotate and stretch S_2
(A.F. Huxley and Simmons, 1971). (From A.F. Huxley, 1974)

the process would not account for the early rapid motion in the contraction transients. Therefore the Huxley-Simmons proposal implies
both a particular force generating process and a time scale for the
kinetics of this process.

I. Stiffness Changes Expected from the Podolsky-Nolan Model

A corollary of the Huxley-Simmons (1971) proposal is that cross-
bridges form at a moderate rather than a fast rate, so that the number
of cross-bridges in the fibre - and therefore the stiffness - would
not be expected to increase in the first few milliseconds after a
quick displacement. Huxley and his colleagues (Ford et al., 1974) have
recently examined this point experimentally, and they were unable to
detect an increase in stiffness. Their experiments have not been described in detail yet, but I thought it would be worthwhile at this
stage to point out the magnitude of the stiffness increase that would
be expected if cross-bridges were formed as rapidly as Nolan and I
have suggested.

Fig. 10 shows calculations made from the model that Nolan and I (1973) put forward. In the upper half of the slide we assumed that $\underline{f}$, the rate function for forming cross-bridges, was 120 Å wide. As $\overline{I}$ mentioned previously, the initial cross-bridge distribution function has the same width. The force-displacement relation, or stiffness, for this initial distribution is given by the lower solid line on the right-hand side; this corresponds to the T_1 curve of Huxley and Simmons (1971). If an exponential compliance were added in series with the cross-bridges, the sarcomere would be less stiff. The upper solid line on the right-hand side shows the overall stiffness when the added series

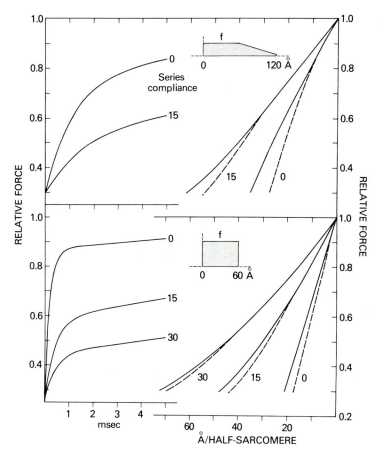

Fig. 10. Stiffness changes expected from the Podolsky-Nolan (1973) model following a step change in length. Curves on the <u>left</u> show the redevelopment of force following a length step that reduces the force from the steady isometric value to about 0.3 times that value. The curves were calculated for an $\underline{f}$ function 120 Å in width (upper inset) and 60 Å in width (lower inset), and for either the cross-bridges alone or with an added exponential compliance. The labels show the displacement, in Å/half-sarcomere, required to reduce the force in the added compliance to half value. Curves on the <u>right</u> show the stiffness of the half sarcomere during the initial length step (solid lines) and 5 msec after the initial length step (interrupted lines). Although the change in stiffness depends on the width of the $\underline{f}$ function, the stiffness is always greater 5 msec after the initial step

compliance is such that a displacement of 15 Å reduces the force to half the initial value.

The traces on the left show the calculated redevelopment of force after a quick displacement, both with and without the added series compliance. After 5 msec, a second quick displacement is applied; the T_1 curve for the second step is given by the interrupted lines on the right. The stiffness of the sarcomere clearly increases in both cases, although the change in stiffness is smaller when a series compliance is present.

The width of the $\underline{f}$ is not uniquely defined by the isotonic velocity transients, and the lower half of Fig. 10 shows corresponding calculations with an $\underline{f}$ function 60 Å in width. The smaller width of $\underline{f}$ is compensated for by a proportional increase in its amplitude. The cross-bridges alone are stiffer in this case, since less displacement is required to produce a given change in relative force.

As shown on the left, force redevelops more quickly with the 60 Å $\underline{f}$; this is due to the greater amplitude of $\underline{f}$. A second quick displacement applied after force has redeveloped for 5 msec produced the interrupted T_1 curves on the right. The T_1 curve that corresponds to the latest data from Huxley's laboratory (Huxley, 1974) is that with the 15 Å series compliance. This series compliance is not unreasonable; it corresponds to a decrease in Z line width of about 2% when the force falls from the isometric values to half this value. In this case, the T_1 curve after 5 msec is less than 15% stiffer than the initial T_1 curve. It will be of interest to see whether the uncertainties in the stiffness measurements, due to factors I have discussed earlier, are small enough to rule out changes of this magnitude.

J. Skinned Fibre Experiments

Another method of studying the cross-bridge mechanism is to see how it is affected by the chemical environment. To do this, the permeability barrier of the surface membrane must be eliminated, without affecting the myofilaments. The first successful experiments along these lines were made by Szent-György (1951), who broke down the surface membrane by extracting fibres with mixtures of glycerol and water. The physiological properties of these preparations were studied extensively in this country be H.H. Weber and his associates (Weber and Portzehl, 1954).

My colleagues and I undertook similar experiments using the skinned muscle preparation worked out by Natori (1954). One advantage of skinned fibres over glycerol-extracted fibres is that removal of the surface membrane by microdissection does not affect the remainder of the fibre. It has also turned out that the contraction kinetics of skinned fibres are highly reproducible. These preparations can be carried through many activity cycles, even when they are allowed to shorten, particularly when the temperature is lowered to around 0°C. The explanation for this is not clear, but it might be that activation is more uniform in skinned fibres than in more drastically treated preparations. The sarcoplasmic reticulum is still functional, and activation probably occurs with a sudden release of calcium from the sarcoplasmic reticulum (Ford and Podolsky, 1972), as in the case of physiological activation.

Our initial studies were concerned with the steady-state behavior of the fibres, and the effect of calcium ions and ionic strength on the force and shortening velocity. Next we quickened the response time of our force and displacement transducers, and Gulati and I have been able to record the contraction transients in skinned fibres. These studies are still in a preliminary state, since the detailed interpretation of the various phases of the contraction transient is still not settled. Nevertheless, even at this stage, we can draw a number of interesting conclusions regarding the influence of various chemical factors on the cross-bridge mechanism.

Fig. 11 shows the response of a fully activated skinned fibre segment to step changes in load (Podolsky et al., 1974). The time scale is 20 msec, and the displacement bar is 50 Å/half sarcomere. The striking thing about these records is that they look exactly like the contraction of an intact fibre at this temperature except that the time scale of the motion is stretched out by a factor of two. The duration of the early fast phase (which ends at α) and of the subsequent slow phase (which ends at τ) are both stretched out to the same extent. This suggested to us that the processes involved in both the fast and the slow phases of the motion are intimately linked, which would necessarily

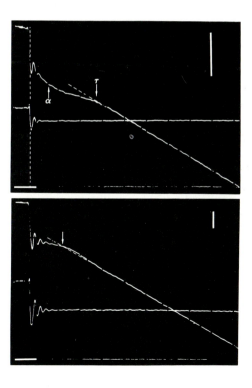

Fig. 11. Response of skinned muscle fibre to step changes in load. Top trace, displacement; lower trace, force. The steady isometric force is 72mg weight; vertical scale bar, 50 Å/half sarcomere; horizontal scale bar, 20 msec. α is the point at which the transient velocity is equal to the final steady velocity; τ is the null time. (From Podolsky et al., 1974)

be the case if attached cross-bridges at a given relative position x
exist in only one kinetically significant state (Hill et al., 1974).
A similar linkage could occur if the cross-bridge existed in more than
a single state, as suggested by Huxley and Simmons (1971), but special
constraints would have to be imposed on the kinetic parameters in this
case.

Similar high time resolution force step experiments were made at different calcium ion levels. This was done because, as discussed earlier,
calcium changes the steady isometric force withouth changing the contraction velocity at a given relative load. If the role of calcium
were limited to an on-off, switch-like activation process, one would
expect the contraction transient at a given relative load to be the
same at all calcium levels. This turned out to be the case experimentally (Gulati, unpublished experiments).

Ionic strength is similar to calcium in the sense that it also modulates the isometric force without changing the steady contraction speed
(Thames et al., 1974). This could mean that KCl, like Ca, affects the
availability of sites at which cross-bridges can be made. However, it
could also mean that ionic strength changes the rate functions for
cross-bridge turnover in such a way that contraction velocity is unaffected, in which case the contraction transients at the same relative
load would be expected to vary with ionic strength.

Fig. 12 shows that the latter alternative is correct (Gulati, unpublished experiments). For the experiment on the left, the physiological
bathing solution (KCl, 5 mM ATP, 1 mM $MgCl_2$, 10 mM imidazole, 5 mM
EGTA, pH 7.0, pCa 5.0) contained 140 mM KCl, while on the right the
KCl concentration was increased to 210 mM. This increase in ionic
strength caused the steady isometric force to drop to half the original

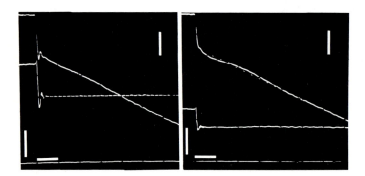

Fig. 12. Influence of ionic strength on the contraction kinetics of
skinned muscle fibres. Top trace, displacement; lower trace, force.
The bathing solution contained 140 mM KCl in the experiment on the
left and 210 mM on the right. Scale bars: displacement, 50 Å/half
sarcomere; force, 20 mg weight; time, 20 msec. Note that increasing
the ionic strength decreases the isometric force, but does not affect
the steady contraction velocity at a given relative force. In addition,
increasing the ionic strength increases the amplitude of the isotonic
velocity transient. (J. Gulati, unpublished experiment)

value. The steady velocities were the same in the two solutions. However, the amplitude of the contraction transient was greater at the higher ionic strength, which shows that the kinetic properties of the cross-bridges in this solution are different from those in 140 mM KCl. We cannot specify the exact nature of these differences until we decipher the language of the transients. However, it is already evident that the effects of known chemical changes on the contraction kinetics of muscle fibres will be a source of new information about the cross-bridge mechanism in the coming years.

Acknowledgment. I am grateful to Dr. Leepo Yu for carrying out some of the calculations shown in Fig. 10, to Dr. Jagdish Gulati for allowing me to include Fig. 12, and to Dr. Terrell Hill for helpful discussion.

References

Civan, M.M., Podolsky, R.J.: J. Physiol. (Lond.) 184, 511-534 (1966).
Ford, L.E., Huxley, A.F., Simmons, R.M.: J. Physiol. (Lond.) 240, 42-43 (1974).
Ford, L.E., Podolsky, R.J.: J. Physiol. (Lond.) 223, 22-33 (1972).
Gordon, A.M., Huxley, A.F., Julian, F.J.: J. Physiol. (Lond.) 184, 170-192 (1966).
Hellam, D.C., Podolsky, R.J.: J. Physiol. (Lond.) 200, 807-819 (1969).
Hill, T.L., Eisenberg, E., Chen, Y., Podolsky, R.J.: Biophys. J. 15, 335-372 (1975).
Huxley, A.F.: Prog. Biophys. 7, 255-318 (1957).
Huxley, A.F.: J. Physiol. (Lond.) 243, 1-43 (1974).
Huxley, A.F., Simmons, R.M.: Nature 233, 533-538 (1971).
Huxley, H.E.: J. Biochem. Biophys. Cytol. 3, 631-648 (1957).
Huxley, H.E.: Science 164, 1356-1366 (1969).
Jewell, B.R., Wilkie, D.R.: J. Physiol. (Lond.) 143, 515-540 (1958).
Natori, R.: Jikeikai Med. J. 1, 119-126 (1954).
Podolsky, R.J.: Nature 188, 666-668 (1960).
Podolsky, R.J.: In: Aspects of Cell Mobility, pp. 87-99. Cambridge University Press 1968.
Podolsky, R.J., Gulati, J., Nolan, A.C.: Proc. Nat. Acad. Sci. USA 71, 1516-1519 (1974).
Podolsky, R.J., Nolan, A.C.: Cold Spring Harbor Symp. Quant. Biol. 37, 661-668 (1973).
Podolsky, R.J., Nolan, A.C., Zaveler, S.A.: Proc. Nat. Acad. Sci. USA 64, 504-511 (1969).
Podolsky, R.J., Teichholz, L.E.: J. Phyiol. (Lond.) 211, 19-35 (1970).
Schoenberg, M., Wells, J.B., Podolsky, R.J.: J. Gen. Physiol. 64, 623-642 (1974).
Sugi, H.: J. Physiol. (Lond.) 225, 237-253 (1972).
Szent-Györgyi, A.: Chemistry of Muscular Contraction, Second Edition. New York: Academic Press 1951.
Thames, M.D., Teichholz, L.E., Podolsky, R.J.: J. Gen. Physiol. 63, 509-530 (1974).
Weber, H.H., Portzehl, H.: Prog. Biophys. 4, 60-111 (1954).

Discussion

Dr. Weber: Should one conclude then that changes of ionic strength alter the behaviour of each bridge whereas Ca only alters the number of bridges?

Dr. Podolsky: Yes, that's a better way of putting it. If I had not been so rushed I might have thought of that nice way of saying it.

Dr. Paul: What do you see as the future steps your laboratory will be undertaking to differentiate between the current cross-bridge kinetic schemes?

Dr. Podolsky: One way is to use some of the X-ray diffraction methods that H. Huxley described to get a measure of the cross-bridge number. Another is to apply force steps to a muscle fibre. But in that case you must assume that the force is uniform along the length of the muscle fibre for a direct interpretation of the experiment, and this is an assumption made by Ford et al. I should say that they have chosen very short muscle fibres and they argue that errors due to the (finite) transmission time are not more than a few percent. But another approach is to consider the transmission time effect as a good thing rather than a bothersome thing, and to actually measure the transmission time itself. This is one of the things we have been trying to do. If you measure the transmission time of the signal down the length of the muscle fibre in different physiological states, you can find out whether the fibres differ in stiffness. Now the advantage of this is that the measurement time in the transmission time method is two orders of magnitude better than in the stepping method. This is because measure the motion of the wave front as it goes along the sarcomere. The sarcomeres in front of the wave don't know the wave is coming, and the sarcomeres in back of the wave don't affect the wave front. So you can measure stiffness with this method in about 10 μsec. Another technique is to use skinned fibres and find out how the fast phase and the slow phase of the transients are related to each other. According to the Huxley-Simmons scheme, there is no necessary connection between the two phases of the motion, whereas according to our scheme there is a connection. One can do various things to see whether the two phases are coupled together or whether they can be decoupled.

Dr. Gergely: Did you imply that under the conditions of the experiments of Ford et al. the two schemes would be experimentally almost indistinguishable?

Dr. Podolsky: I am saying that in our scheme you always expect an increase in stiffness and that depending on the specific rate constants this increase in stiffness may become pretty small. In the graph I showed it was a change of 15%. To exclude this their experiments have to be better than 15%.

Energy Transformation in Muscle

D. R. Wilkie

I wish to begin by paying tribute to Professor H.H. Weber both as a scientist and as a warm-hearted human being.

The saddest thing of all is that he cannot be with us at this meeting, where he must certainly have felt great satisfaction at the way in which his own life of research has influenced the work of all of us here. More than anyone else he persuaded biochemists to learn physiology and - I speak from my own experience - his work convinced physiologists that they must learn biochemistry.

A. Introduction

The contractile filaments of muscle form a machine made of protein in which a particular chemical reaction, the hydrolysis of ATP, proceeds in such a way that it leads to the production of mechanical work. This provides a most dramatic instance of energy transformation. Muscles compare quite favourably with man-made machines (Wilkie, 1969) both as generators of force (2-4 atmospheres) and of mechanical power (0.3 HP/kg during a single contraction (Wilkie, 1960a); about 1/5 that produced continuously by petrol engines) (Wilkie, 1959). Since we are concerned with chemical change, heat and work, we are brought face to face with the problems of thermodynamics. From the formal thermodynamic point of view, muscle exactly corresponds to a galvanic cell connected to an electric motor, in which similarly a chemical process occurs in such a way that work is performed.

Now, although muscles are deplorably complicated, they (along with many other biological systems) do at least possess several features that simplify the application of thermodynamics. They are for all practical purposes isothermal systems; the temperature gradients that inevitably arise in them are small and obviously incidental to the way in which the muscle functions. Also, since they form a condensed system in which the chemical reactions proceed in solution, volume changes are negligibly small, about one part in 10^5, so the pressure-volume work done against the atmosphere is always negligible. The difference between changes of energy and of enthalpy is thus also negligible and the difference between Helmholtz' and Gibbs' free energy can likewise be disregarded. By suitable experimental design, we can arrange that the protein contractile machinery and its substrates function either as closed systems (i.e. exchanging energy only with their surroundings) or as open ones (which exchange matter as well as energy with their surroundings). I have given elsewhere an account of the relevant thermodynamic arguments (see Wilkie, 1960b, 1967, 1970, 1974) which I will not reproduce here. For our present purposes it will suffice to give two simple equations, the first (1) derived from the First Law;

$$\underline{h} + \underline{w} = -\Delta \underline{U} \simeq -\Delta \underline{H} \tag{1}$$

$\underline{h}$ is the heat produced, $\underline{w}$ is the work produced, $\Delta\underline{H}$ is the change in enthalpy and $\Delta\underline{U}$ is the change in energy.

The second equation (2) incorporates the Second Law and tells us that to each specified change there corresponds a maximum value of the work that can be produced ($\underline{w}_{max}$); $\Delta\underline{S}$ is the change in entropy resulting from the process,

$$\underline{w}_{max} = -(\Delta\underline{U} - \underline{T}\Delta\underline{S}) = -\Delta\underline{A} \simeq -\Delta\underline{G} \tag{2}$$

$\underline{T}$ is the absolute temperature, $\Delta\underline{A}$ is the change in Helmholtz' free energy and $\Delta\underline{G}$ is the change in Gibbs' free energy. It will be appreciated that in a chemical reaction, the products usually differ both in energy $\underline{U}$ and in entropy $\underline{S}$ from the reactants. Eqs. (1) and (2) apply to a closed but not isolated system under the isothermal conditions in which most experiments on muscle have been conducted. In what follows I shall use the functions $\Delta\underline{H}$ and $\Delta\underline{G}$ because they are probably more familiar and marginally more accurate than $\Delta\underline{U}$ and $\Delta\underline{A}$.

If only one chemical reaction were proceeding in the muscle, we could write Eq. (1) in the form (3)

$$(\underline{h} + \underline{w}) = -\Delta\underline{H} = -\Delta\xi\Delta\underline{H}_m \tag{3}$$

where $\Delta\underline{H}_m$ is the molar enthalpy change of the reaction (in J/mol) and $\Delta\xi$ (in mol) is the extent of reaction during the time interval in question. All the four terms of this equation are determinable experimentally; $\underline{h}$ with sensitive thermopiles, $\underline{w}$ with suitable mechanical levers and transducers, $\Delta\underline{H}_m$ by independent calorimetric investigations and $\Delta\xi$ by chemical analysis of the contracting muscle. The determination of $\Delta\xi$ presents the problem that since chemical analysis is a destructive process one cannot analyse the self-same muscle both before and after it has contracted. It is, therefore, necessary to conduct such experiments in duplicate with one muscle as an unstimulated control while the other muscle is stimulated and thus caused to contract in whatever way is under investigation. Both muscles are otherwise treated identically and are rapidly frozen within about 60 ms in a special 'hammer' apparatus (Kretzschmar and Wilkie, 1969) in which the muscles are flattened between metal surfaces previously chilled to -196°C. The muscles can subsequently be extracted with perchloric acid and analysed for substrates of interest. Recent improvements have been made in the sensitivity of NMR spectroscopy as applied to the ^{31}P-containing compounds that interest us in muscle (Hoult et al., 1974). The sensitivity of this method still seems to be very low compared with that of conventional chemical analysis. However, the possibility of making chemical analyses by a physical technique on intact living muscle is so attractive that this technique deserves close attention.

If all the terms in Eq. (3) are known, what is the point of the exercise? The answer is that the physical measurements of $\underline{w}$ and $\underline{h}$ cross-check the calorimetric and chemical determinations. If Eq. (3) is found experimentally not to be obeyed (and this is what has been found in all the experiments conducted until now) then we know that our knowledge of the chemical processes must be incomplete, an item of information that cannot be obtained in any other way. Each additional reaction will contribute its own term similar to the right-hand side of Eq. (3).

B. Experimental Results

Some years ago the situation appeared to be relatively simple, as shown in Fig. 1, at least for muscles poisoned with iodoacetate and nitrogen. Biochemical evidence abounds indicating that the contractile system of

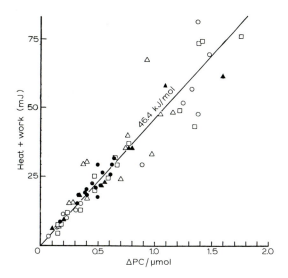

Fig. 1. The relation between heat, work and splitting of phosphocreatine (PC) in various types of contraction including relaxation and an additional delay of about 45 s. Muscles at $0^{\circ}C$ treated with iodoacetic acid and nitrogen. Ordinate: (heat produced + work produced) by muscle, equal to the negative change in enthalpy of muscle, in mJ. Abscissa; amount of phosphocreatine split (ΔPC) in µmol: O from 4 to 107 isometric twitches, 12 experiments; ● 30 isometric twitches, 15 experiments; □ isometric tetani lasting from 7 to 111 s, 16 experiments; ▲ from 8 to 78 isotonic twitches, with performance of positive mechanical work, 9 experiments; △ from 2 to 8 tetani, duration 7 s, with slow isotonic stretch and negative work, 13 experiments. The line has been drawn with a slope of 46.4 kJ/mol (as explained in the text). (From Wilkie, 1968)

such muscles forms a closed system in which the only reaction to be expected is the hydrolysis of phosphocreatine. The experiments were conducted on a slow time scale: each pattern of contraction included the phase of relaxation and usually a fair amount of time elapsed between contractions. Since the 'hammer' apparatus had not yet been invented there was also an interval of some 45 s between the end of the last relaxation and the moment of freezing.

On this slow time scale, it is evident that there is fair proportionality between the output of energy (i.e. $\underline{h}$ + $\underline{w}$) and the breakdown of phosphocreatine. The slope of the line indicates that one is obtaining some 46 kJ/mol of phosphocreatine split. At the time (Wilkie, 1968), this was close to the accepted calorimetric value for the molar enthalpy change of phosphocreatine splitting so it appeared that Eq. (3) was, under these circumstances, being obeyed and as though only one reaction was occurring in the muscle.

Since that time, various discrepancies have appeared which are still
not resolved. In the first place, careful calorimetric determinations
(Woledge, 1972) indicate that one can expect only 34 kJ/mol for the
hydrolysis of phosphocreatine and for the subsequent reactions with
the buffers thought to be present in muscle. Secondly, technical improvements of various kinds have made it possible to examine energy
balance in contracting muscle on a fast time scale and to follow the
output of energy ($\underline{h}$ + $\underline{w}$) and the chemical changes from instant to instant during a single maintained contraction. From Fig. 2 it is evident
that early in contraction a large amount of heat appears that cannot
be accounted for by concurrent splitting of phosphocreatine. The chemical and physical measurements have been matched with the old value
of 46 kJ/mol rather than the new one of 34 kJ/mol. If the latter value
is adopted the discrepancy becomes even greater. The obvious inference
is that in the early stages of contraction there must be an unidenti-
fied exothermic process which we have reason to think may be reversed
during a period lasting several minutes after relaxation has ended,
as shown in Fig. 3. This effect was described many years ago by Lundsgaard (1934).

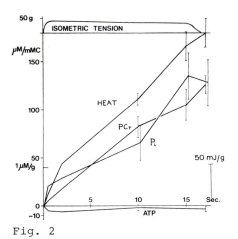

Fig. 2

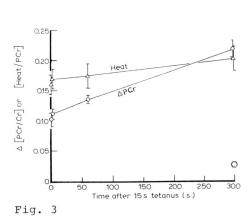

Fig. 3

Fig. 2. The relation between heat produced and chemical change during
a single isometric tetanus of 15 s duration at $0^{o}C$; normal muscles in
oxygen. Breakdown of phosphocreatine (PCr) and ATP, and liberation of
inorganic phosphate are shown as upward deflections. Heat production
has been scaled at 46.4 kJ/mol, as in Fig. 1. The upper trace shows a
record of tension development and relaxation. (From Gilbert et al.,
1971)

Fig. 3. Post-contractile splitting of phosphocreatine (PCr). The experiment has been preceded by a 15 s tetanus similar to that in Fig. 2.
The muscles, at $0^{o}C$, were treated with iodoacetic acid and nitrogen
to prevent recovery resulting from glycolysis or oxidative phosphorylation. As in Fig. 2, heat and PCr have been scaled at 46.4 kJ/mol. The
hexagon indicates that even after 300 s extremely little fructose 1,6-
diphosphate has been formed. (From Kretzschmar and Wilkie, unpublished
results)

We have found that heat production exceeds phosphocreatine breakdown
in the first second or two of contraction consistently in our experi-

ments on English frogs *(Rana temporaria)*. However, Homsher et al. (1975), who have made similar experiments on the American frog *(R. pipiens)*, find no obvious energy discrepancy early in contractions such as is shown in Fig. 2 though they do obtain 46 kJ/mol of phosphocreatine breakdown, so their results remain inconsistent with calorimetric determinations. With *R. temporaria* they obtain results virtually identical with our own. Clearly, this situation demands further experimental investigation. If indeed it can be confirmed that there is no energy discrepancy in *R. pipiens* this might provide the explanation for the otherwise puzzling finding by Marechal and Mommaerts (1963) that in *R. pipiens* no post-contractile PCr splitting could be detected.

C. Work Balance

From the thermodynamic point of view, work and heat are very different types of energy. One inevitable question about the contractile machine is, how efficiently does it transduce chemical into mechanical energy? Referring to Eq. (2) we can compare the work observed experimentally ($\underline{w}$) with the maximum that might theoretically be expected ($\underline{w}_{max}$), and derive Eq. (4) [analogous the Eq. (3)]; $dG/d\xi$ is often designated by $\Delta\underline{G}$ or $\Delta\underline{G}_m$ (Guggenheim, 1967) so care must be taken to avoid confusion.

$$\underline{w} \leqslant \underline{w}_{max} = -\Delta\underline{G} = \int_{\xi_2}^{\xi_2} \frac{dG}{d\xi} d\xi \tag{4}$$

Integration is necessary because $dG/d\xi$ alters as the reaction proceeds. Unfortunately, $dG/d\xi$ cannot be measured directly at present; it can only be calculated and for this pupose we have used the results of Alberty (1972).

In the actual experiment (Curtin et al., 1974), we compared two brief contractions. In one of these we held the muscle at constant length so that its performance of work was minimized, whereas in the other we allowed the muscle to shorten at the speed at which it generates maximal mechanical power. The duration of the non-working contraction (IM) was almost doubt that of the working (ERG) contraction. We did this to make the total energy output similar in both cases; thus, in one contraction almost all the energy came out as heat whereas in the other about 35% came out as work. The effect on phosphocreatine breakdown is dramatic: it is almost doubled in the working contraction and so the rate of breakdown is practically trebled, see Table 1. In optimal conditions we obtained about 26 mJ of work/g muscle and the corresponding (calculated) value of ΔG was about 60 mJ/g muscle. Although heat may be produced by an unknown source, we thus have no reason to suppose that mechanical work comes elsewise than from concurrent chemical change. Taking the simplest view, that mechanical work is indeed derived from concurrent chemical change, the efficiency of transduction thus appears to be only about 50%. Experiments on this topic (and with *R. pipiens*) were reported several years ago by Kushmerick and Davies (1969), who found higher values of efficiency than we have done (up to 0.66). In their experiments, as in ours, there was always sufficient chemical change to account for work produced. Incidentally, the experiments by Curtin et al. (1974) provided further confirmation of the results shown in Fig. 2. Phosphocreatine splitting sufficed to account for the work observed, but not for the total energy observed: a considerable excess of heat production was always found.

Quite apart from questions of energy and work balance, the direct
measurements of rates of phosphocreatine splitting have some interest
of their own is showing quantitatively the extent to which the primary
chemical process in the cross-bridges, the hydrolysis of ATP, is controlled by mechanical conditions and by the low calcium level that is
believed, on good grounds, to exist in resting muscle.

Table 1. Splitting of phosphocreatine (or ATP) in frog sartorius (*R.
temporaria*) at 0°C. (From Curtin et al., 1974)

	Rate of Splitting	Cross-bridge Mean Lifetime
	mmol kg^{-1}sec^{-1}	sec
Isometric	0.4	⩽0.35
Maximal working	1.12	⩽0.12
Resting	<0.5 × 10^{-3}	>200

The results are shown in Table 1. The chemical change shown is the
breakdown of PCr rather than of ATP but there is abundant evidence to
show that in living muscle the Lohmann reaction is both complete and
rapid, so ΔATP = ΔPCr. The left-hand column shows the striking effect
of work performance in practically trebling the rate of chemical breakdown.

The turnover of ATP in the cross-bridges of resting muscle cannot be
determined directly by experiment, but an upper limit can be set to
it by considering the resting oxygen consuption, since resting muscle
is in a steady state in which its levels of ATP and of PCr are altering
only slightly, if at all. We may make the extreme assumption that all
the resting O_2 consumption is devoted to rebuilding the ATP that has
been accidentally broken down as a result of imperfect control. Taking
direct determinations of O_2 consumption at 0°C (7.3 × 10^{-9} mol min^{-1}g^{-1})
and of *in vivo* P:O ratio (2:1) from Kushmerick and Paul (1974), it is
easy to calculate that the rate of ATP breakdown at rest must be less,
perhaps much less, than 0.5 × 10^{-3} mmol sec^{-1}g^{-1}. The control ratio
(the ratio between the maximum rate of ATP splitting and the minimum
rate) must therefore be greater than 1.12/0.5 × 10^{-3} = 2,200:1. This
is much greater than can be achieved in *in vitro* preparations of actomyosin and sarcoplasmic reticulum. Perhaps a dual mechanism involving
myosin as well as troponin-actin is needed in order to achieve such
a large degree of control.

One may also make some calculations that at least set limits for some
of the events at the cross-bridges - limits that will be refined as
our knowledge of the cross-bridges themselves improves.

We may take the myosin concentration in muscle as 0.14 mmol kg^{-1} (see
Ebashi et al., 1969) and assume that in each cross-bridge cycle one
ATP molecule is hydrolysed. There is at present no evidence to suggest
that two ATP molecules must be split, corresponding to the two S_1 units
of myosin. Under isometric conditions the mean duration of a cross-
bridge cycle will be 0.14/0.4 = 0.35 sec if all the cross-bridges
participate, less than 0.35 sec if only of a fraction of them do so.
The calculation for the working contraction is made in a similar way.
In the resting muscle there is no reason to suppose that only a fraction of the cross-bridges escape from control, so the inequality sign
can be put as shown in the Table.

Our knowledge of the precise arrangement of the cross-bridges on the
thick filament remains somewhat uncertain and we do not, of course,
know to what degree they can move in the equatorial plane so as to
engage either of the two nearest-neighbour thin filaments.

However, by making what at the moment seem to be the most reasonable
assumptions about these questions one can arrive at a maximum estimate
of approximately 40 nm for the spacing of the sites on the thin filament with which a given cross-bridge can interact (for further details
see Curtin et al., 1974). This estimate is close to the repeat distance
along the actin helix. To the extent that it can be depended on, this
result suggests that although every actin monomer is a potential site
for interaction with myosin (as shown by the experiments in which thin
filaments are decorated with S_1) within the actual muscle the local
architecture dictates that only the actin monomer directly facing a
cross-bridge may be able to interact with it.

I wish to end by reporting some very recent work from our laboratory.

Kodama and Woledge have measured the heat of binding ADP to highly-
purified myosin in a modified LKB microcalorimeter. A great deal of
information can be obtained from this type of measurement. Adding
successive equal aliquots of ADP and measuring the resulting equal in-
crements in heat production yields the value 54 kJ $(mol\ ADP)^{-1}$ at $0°C$
in $0.5\ mol\ 1^{-1}$ KCl, pH 7.8. Beyond a certain point the myosin is saturated with ADP and no further heat is produced. Identification of this
point reveals the stoichiometry of the reaction, which comes out at
1.9 mol ADP per mol myosin. When the myosin is almost, but not completely, saturated with ADP, the reactin is incomplete. From the cur-
vature of the relation between heat produced and ADP added, the binding
constant may be calculated. It comes out at $10^{6.2}$.

Clearly this technique has great potential for studying other steps in
the kinetic sequence of events at the cross-bridges.

We are also looking actively into the possibility of using the improved
^{31}P NMR technique developed in the Biochemistry Department of Oxford
(Hoult et al., 1974) in order to study living muscle. The physiological
problem here is to develop methods of keeping muscles alive, of stimulating them and if possible recording their tension development in
the extremely small space available within the superconducting magnet
of the instrument without upsetting the finely adjusted homogenous
magnetic field or the tuning of the circuits. Our preliminary experiments lead us to feel optimistic that these problems can be solved.

References

Alberty, R.A.: In: Horizons of Bioenergetics, pp. 135-144. New York:
  Academic Press 1972.
Curtin, N.A., Gilbert, C., Kretzschmar, K.M., Wilkie, D.R.: J. Physiol.
  (Lond.) 238, 455-472 (1974).
Ebashi, S., Endo, M., Ohtsuki, I.: Q. Rev. Biophys. 2, 351-384 (1969).
Gilbert, C., Kretzschmar, K.M., Wilkie, D.R., Woledge, R.C.: J. Physiol.
  (Lond.) 218, 163-193 (1971).
Guggenheim, E.A.: Thermodynamics, an Adwanced Treatment for Chemists
  and Physicists. Amsterdam: North Holland 1967.
Homsher, E., Rall, J.A., Wallner, A., Ricchutti, N.V.: J. Gen. Physiol.
  65, 1-21 (1975).

Hoult, D.G., Bushby, S.J.W., Gadian, D.G., Radda, G.K., Richards, R. E., Seeley, P.J.: Nature 252, 285-287 (1974).

Kretzschmar, K.M., Wilkie, D.R.: J. Physiol. (Lond.) 202, 66-67 (1969).

Kushmerick, M.J., Davies, R.E.: J. Proc. R. Soc. Lond. B. 174, 315-353 (1969).

Kushmerick, M.J., Paul, R.J.: Biochim. Biophys. Acta. 347, 483-490 (1974).

Lundsgaard, E.: Biochem. Z. 269, 308-328 (1934).

Marechal, G., Mommaerts, W.F.H.M.: Biochim. Biophys. Acta 70, 53-67 (1963).

Wilkie, D.R.: Nature 183, 1515-1516 (1959).

Wilkie, D.R.: Ergonomics 3, 1-8 (1960a).

Wilkie, D.R.: Prog. Biophys. Biophys. Chem. 64, 471-481 (1960b).

Wilkie, D.R.: Symp. Biol. Hung. 8, 207-235 (1967).

Wilkie, D.R.: J. Physiol. (Lond.) 195, 157-183 (1968).

Wilkie, D.R.: Nature 221, 306 (1969).

Wilkie, D.R.: Chem. Br. 6, 472-476 (1970).

Wilkie, D.R.: J. Mechanochem. Cell Motility 2, 257-267 (1974).

Woledge, R.C.: Cold Spring Harbor Symp. on Quant. Biol. 27, 629-634 (1972).

Discussion

Dr. Podolsky: What is the definition of "mean life time" in a "maximally working muscle", and how does this parameter relate to the "turn-over time" of a cross-bridge, as estimated from the 10 msec duration of the isotonic velocity transient?

Dr. Wilkie: All I have done is to divide the myosin content, which is approximately 0.14 mol kg^{-1} by the observed rate of ATP splitting in mol kg^{-1} sec^{-1}, so you end up with a time in sec. These calculations are explained in Curtin et al. (1974), p. 468-369.

Dr. Podolsky: You could take those two numbers and draw a conclusion.

Dr. Wilkie: Well I don't see how you could draw that conclusion without any additional evidence. What I have done is to present these calculations as setting extreme limits. Similar limit-case calculations make it possible to say what is the furthest apart that the sites on the thin filament can be: this comes out at 44 nm.

Dr. Hasselbach: Has the presence of myokinase been excluded in the experiments by Kodama and Woledge for measuring the binding heat of ADP?

Dr. Wilkie: Oh yes, in this experiment one of the main difficulties is to get rid of myokinase. If myokinase is present the whole thing is invalid because the heat of binding is mixed with heat of splitting of ATP that is continually formed. Over the whole duration of the experiment only 3% of the ADP was split. Certainly this is a major experimental difficulty, which they overcome by taking great pains over the purification of their myosin.

Dr. R. Paul: Recent work of Homsher shows no "energy gap" of the first type in American frogs (*Rana pipiens*), would you comment on how you think this may influence your interpretation of this "gap".

Dr. Wilkie: What this relates to is a recent paper by Homsher and his colleagues that some of you may have seen in the Journal of General

Physiology 65, 1-21 (1975) in which he repeated experiments like the one I showed in my second slide. In *Rana pipiens*, they didn't find a discrepancy early in contraction as we had found, though in *Rana temporaria* their results were exactly like ours. Even in *R. pipiens* there is still a disagreement with calorimetric findings since they obtain about 46 kJ of energy per mol of phosphocreatine split. If true, this would relegate the extra heat that appears in our experiments to a function other than being directly involved with the contractile machinery. For example, it might be something a bit on one side connected with differences in activation.

If true, it would certainly resolve some other discrepancies; for example, the fact that European frogs have, ever since the days of Lundsgaard, been known to show the phenomenon of post-contractile phosphocreatine splitting seen in my slide 3. After contraction and relaxation are over, European frogs go on splitting phosphocreatine, which of course they would need to do because finally they have to end up in balance. We have confirmed this in our experiments on *R. temporaria*, whereas there is an apparently very solid-looking paper from Mommaerts' laboratory working with *R. pipiens* about 1962 saying that there is no post-contractile phosphocreatine splitting. On the experimental side, which of course is where things will finally be resolved, what is satisfactory is that Homsher imported some *R. temporaria* into Los Angeles and found exactly the same as we have found. We have done the converse experiment of importing some *R. pipiens* into England and doing the same experiment and I had hoped to bring the full results with me. All that I can say at the moment is that the chemical change in these muscles is exactly the same as in *R. temporaria* and very similar to what was reported by Homsher. We are not yet quite sure about the heat measurements because our co-worker, Merlin Kretzschmar, has gone off to San Francisco and some vital bits of information I needed to calibrate the heat records are in the TransAtlantic mail. Nevertheless we should have an answer fairly soon.

Added in proof: The heat measurements have now been completed. In brief, we do not confirm the results of Homsher et al. We find *R. pipiens* to be essentially similar to *R. temporaria*. The results will be communicated to the Physiological Society at its September Meeting, 1975.

Dr. Hasselbach: Does your more recent work confirm the production of ATP in the early phase of contraction?

Dr. Wilkie: Well I don't know because we haven't repeated that particular experiment. The thing I did not mention in connection with slide 2 is that at no time is there any sign of net ATP breakdown and early on in contraction there is actually a slight increase in the ATP level. However you have to do very many experiments, 70 or more, in order to achieve statistical significance. We had 70 experiments done for other purposes and we have never had 70 experiments in another set so we can not say whether we have further confirmed the result or not. However, an increase in ATP now seems less intrinsically unlikely than it did when we found it. We thought it was a mistake but now there is a fair bit of biochemical evidence that the first change after activation may be discharge of ADP into the environment and its rapid phosphorylation. So it does not seem as weird as it did.

Dr. R.J. Paul: I would just like to add to that point that in American frogs (*R. pipiens*), Martin Kushmerick and I have observed exactly the opposite effect. In short isometric contractions we observe an ATP breakdown, in 20 experiments at 1 sec we saw a statistically significant breakdown of 0.3 µmole/g. With tissue-destructive chemical measurements, as you know, the results tend to be quite dependent on the

batch of frogs. With one batch, one finds a small synthesis and with
the next, breakdown; I would not put a great deal of credence on such
small chemical changes. These chemical measurements can be quite frus-
trating and I was interested in your talking about physical measurements to resolve some of these problems.

Dr. Wilkie: Well certainly one is bothered by the statistical nature
of the experiments themselves. However, using physical measurements
like NMR there seems to be absolutely no hope of their being within
a thousandfold as sensitive as the chemical methods. I believe that
they will have other advantages but certainly sensitivity is not one
of them.

Ca^{2+} Action and the Regulation of the Actomyosin ATPase

Release and Uptake of Calcium by the Sarcoplasmic Reticulum

W. Hasselbach

A. Introduction

Today it seems well established that calcium ions are the mediators of excitation contraction coupling. The sequence of the reactions by which muscle activity is twitched on starts with the depolarization of the plasma membrane. Depolarization spreads inwards actively along its narrow transverse tubular invaginations and reaches the innermost fibrils in approximately 1 msec. (Gonzales-Serrates, 1971; Benzanilla et al., 1972; Adrian and Peachey, 1973). There follows a sudden rise of the calcium concentration in the myoplasma and the contractile protein is activated (Ashley and Ridgway, 1970; Rüdel and Taylor, 1973). The muscle relaxes when the calcium concentration has returned to its resting level. Fig. 1 illustrates these events schematically. However,

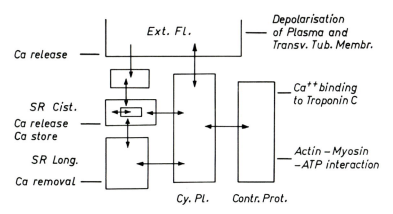

Fig. 1. Schematic diagram of excitation-contraction coupling. The arrows represent active and passive calcium movements

a quantitative evaluation of this diagram is not possible. Neither the size of the various compartments for calcium nor the rate constant describing the calcium movement between the different compartments is known accurately. However, as incomplete this scheme may appear if its quantitative essence is considered, it represents considerable progress. It started with Heilbrunn's observation (Heilbrunn, 1940) that contractions of muscle fibers can be induced by the injection of solutions containing calcium ions and was extended by a series of findings by morphologists, biochemists and physiologists.

1. Between 1957 and 1960 the existence of complicated internal membrane systems in muscle fiber was confirmed (Porter and Palade, 1957). Such structures have been seen before, but were considered to be artefacts.

2. A number of hitherto unexplained properties of the isolated contractile proteins were understood after A. Weber and Winicur (1961)
and Weber and Herz (1963) found that the contractile system requires
low concentrations of ionized calcium for contractile and enzymatic
activity. In the meantime this calcium requirement was recognized to
be the result of a system of regulatory proteins whose interaction
with the contractile proteins depends on the calcium level (Ebashi
and Endo, 1968; Schaub and Perry, 1969; Perry et al., 1972; Potter
et al., 1974).

3. An ATP-driven calcium transport system was discovered in a membrane
fraction of the muscle which later was identified as fragments of the
sarcoplasmic reticulum. The membrane system proves to be able to establish a calcium level as it is required for the function of the regulatory protein system (Hasselbach and Makinose, 1961, 1963; Makinose and
Hasselbach, 1965; Weber et al., 1966).

4. Finally, physiologists have demonstrated that the calcium movements
in the muscle fiber are controlled by the potential across the plasma
membrane and that this change is transmitted by the elaborated system
of the transverse tubular membranes to the innermost fibril (A.F. Huxley, 1959; Adrian and Peachey, 1973).

In the following I want to concentrate on the function of the sarcoplasmic reticulum as far as it is involved in the regulation of muscular activity. Two problems will be discussed.

1. The calcium level and its maintenance in the resting muscle.

2. The liberation and the removal of calcium in the active muscle.

I. The Calcium Level in the Resting Muscle

There is general agreement concerning the threshold concentration of
calcium required for the initiation of contraction. It was first determined by Portzehl et al. (1964) by injecting calcium buffers into
the thick fibers of *Maja squinado*. These concentations found necessary
to activate the living fiber are in excellent agreement with those
required for the activation of isolated contractile structures (Table
1). The observed threshold concentrations are identical for all muscles and amount to 0.4 µM. This concentration is the upper limit of
the free calcium concentration in the resting muscle. In fact, as shown
by Hagiwara and Nakajima (1966) the resting calcium level in the bar-
nacle muscle is considerably lower than the mechanical threshold. By
measuring the equilibrium potential for calcium the level has been
found to be 0.05 µM (Keynes et al., 1973). This concentration is approximately 10,000 times lower than the concentration of the ionized
calcium in the extracellular space and in the lumen of the sarcoplasmic
reticulum. The mechanism by which the high electrochemical potential
is maintained includes a number of intriguing problems. Since during
rest and activity calcium invades the muscle fiber, calcium must be
extruded from the myoplasma permanently. Three mechanisms are under
discussion.

1. The secretion of calcium from the cisternae of the sarcoplasmic
reticulum through the T-tubules has been proposed by Winegrad (1968).

2. A sodium-calcium exchange mechanism slowly moves calcium from the
myoplasma into the extracellular space (Beeler and Reuter, 1970).

3. An ATP-driven calcium pump in the plasma membrane has repeatedly
been proposed.

Table 1. Level of ionized calcium

Resting muscle	Active muscle	
	Mech. threshold	Max. activity
µM	µM	µM
0.08 < C < 0.2 Barnacle (Hagiwara and Nakajima, 1966)	0.3 Frog; Tension	1.0 (Hellam and Podolsky, 1969)
	0.4 Maia; Tension	1.5 (Portzehl et al., 1964)
0.05 Barnacle (Keynes et al., 1973)	0.5 Rabbit; Glyc. Fiber, Tension	1.0 (Portzehl, 1965)
	1 Frog Glyc. Fiber, Tension	10 (Julian, 1971)

Table 2. Resting calcium level and its maintenance by the SR calcium pump

	Ca_o^{++} µM	Ca_i^{++} µM	Ca_i/Ca_o
in vivo	0.05	100?	2,000
Isolated SR	0.002 0.004	6 120	3,000 (Ox) 30,000 (Phosphate)

Table 3. Perturbation of the Ca SR transport under steady-state conditions

Conditions		Ca_i^{++} nM	Ca-Exchange $pmol \cdot cm^{-2} \cdot min^{-1}$
Resting muscle frog 20°C		~50	< 10
Temperature (Ca_i = 4 µM)	20°C 30°C	2 2	0.25 1.1
Jonic strength (Ca_i = 0.2 mM) 30°C	0.24 1.2	10 16	1.5 2.5
Hypertonicity (Ca_i = 4 µM) 30°C	0.2 M 1 M	4.2 2.5	1.0 0.5
Urea (Ca_i = 4 µM) 20°C	0 1.5 M	2 20	0.2 0.6

The participation of the sarcoplasmic calcium pump in the process of calcium elimination is supported by two lines of evidence.

1. In the membranes of the sarcoplasmic reticulum a very active calcium pump is located which can reduce the free calcium concentration far

below the threshold for contraction (Hasselbach and Makinose, 1963; Weber et al., 1966) (Tables 2 and 3).

2. Winegrad (1968) observed that after a period of muscle activity calcium is shifted slowly in the longitudinal elements of the sarco-plasmic reticulum to its cisternal elements. He assumed that the calcium accumulated in the cisternae is excreted into the transverse tubular system. Such a mechanism, however, seems to be not very prob-able because calcium secretion would require that the calcium concen-tration inside the reticulum has to be raised above its value in the extracellular space. This concentration, however, causes a severe in-hibition of the calcium transport mechanism (Makinose and Hasselbach, 1965; Weber, 1971; Ikemoto, 1974) (Fig. 2). This inhibiting effect of

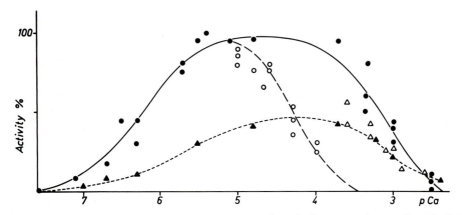

Fig. 2. Activation and inhibition of calcium uptake and calcium-de-pendent ATPase activity by calcium. •-• Calcium-dependent ATPase ac-tivity as observed when the calcium concentration indicated on the abscissa prevails at the external and the internal surface of the sarcoplasmic membranes (ether-treated vesicles). The activity profile between pCa 7 and pCa 5 is identical with that for the calcium-depen-dent ATPase and the calcium transport when the internal calcium con-centration is kept constant by oxalate. o-o, Δ-Δ Calcium uptake activ-ity of native vesicles. The external calcium concentration was kept constant at pCa 5.5. Internal calcium was varied by changing either the oxalate concentration o-o, or the phosphate concentration of the assay medium Δ-Δ. ▲-▲ ATPase activity of closed vesicles in the absence of calcium-precipitating agents

calcium occurs when the internal binding sites of the sarcoplasmic reticulum are occupied by calcium. This severe difficulty argues in favour of the two other proposed mechanisms. The transport of calcium across the external membrane using a sodium-calcium exchange mechanism has been established in heart muscle and in nerve fibers (Beeler and Reuter, 1970; Baker et al., 1969). The energy needed for this kind of calcium elimination is furnished by the sodium potential and conse-quently is provided by the sodium-potassium pump which maintains the sodium potential. There is an interesting relationship between the two calcium eliminating pump systems.

1. The calcium pump of the sarcoplasmic reticulum on the one hand, and the sodium-potassium pump on the other hand, are fuelled by the same energy donor, ATP.

2. Both systems work reversibly.

3. In the resting muscle the electrochemical potential of the ions involved is not far from the chemical potential of ATP. Therefore, the following relationship exists:

$$\frac{Na^3_o \cdot K^2_m}{Na^3_m \cdot K^2_o} \cdot \exp \left(\frac{\Delta V \cdot F}{RT}\right)^2 = \frac{ATP}{ADP \cdot P} \cdot K = \left(\frac{Ca_i}{Ca_m}\right)^2$$

Subscripts m, o, i indicate the presence of the ions in the sarcoplasmic reticulum, the extracellular and myoplasmic space, respectively. [2]The expression takes care of the electrogenicity of the Na-K pump.

The third proposed calcium eliminating mechanism, an ATP-driven calcium pump in the plasma membrane, has been found in erythrocytes (Schatzmann, 1973). Its existence in the plasma membrane of muscle cells, however, is difficult to establish.

II. Calcium Release

Our second subject concerns the release of calcium during excitation. How does the concentration of ionized calcium change when muscle activity is turned on and what are the amounts of calcium shifted during the activity cycle, and from which stores is activator calcium liberated?

There is fair agreement between the experiments performed with living muscle and with isolated contractile structures that the free calcium concentration must be elevated to 1 µM in order to reach maximum activity of the contractile system (Table 1). This calcium concentration, like the threshold concentration, has been measured by the application of calcium buffers. They were injected into living fibers and used in assay media in which the properties of the isolated contractile proteins were analysed. The time course of the calcium transients in the living muscle was monitored by the light emission of the calcium indicator aequorine which has even been injected into the thin toad fibers (Rüdel and Taylor, 1973) (Fig. 3). However, neither the injection of calcium buffers nor the aequorine method gives reliable information concerning the amount of calcium which is necessary to activate the contractile protein. The estimates of these amounts are based on calcium binding studies performed with the contractile protein complex or its isolated components (Table 4). These estimates have furnished continuously increasing values in the last few years from approximately 0.05 µmol ml^{-1} muscle fiber to 0.2 µmol ml^{-1}. The latter figure results if we assume that the troponin C molecule must combine with two calcium ions in order to relieve inhibition of the contractile ATPase completely (Potter et al., 1974). This estimate represents, presumably, an upper limit. If such a requirement of calcium really existed, the muscle would have to mobilize approximately 10% of its total calcium content during each contraction.

Which are the stores that can supply the contractile system with these quantities of calcium? In skeletal muscle, whose fibers have a diameter of approximately 50 µm, the extracellular space cannot serve as calcium store, as has been pointed out by A.V. Hill (1948). Hill argued that the latency period is considerably shorter than the time required for supplying the innermost fibrils with an activator released at the surface membrane by diffusion. In skeletal muscle the internal sarcoplasmic membrane system is the only source which can be considered. If all calcium present in the muscle were concentrated in the sarcoplasmic

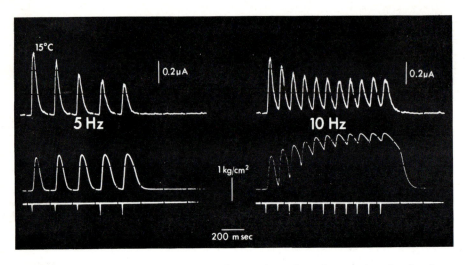

Fig. 3. Calcium transients and tension development of single muscle fibers of the toad (Taylor, 1974). The muscle fibers were stimulated with a frequency of 5 and 10 H_2, respectively. Light emission of aequorine is shown by the upper traces and tension development by the lower traces

Table 4. Ca^{++} binding sites in skeletal muscle

	Conc. of binding sites mmol · 1^{-1}	Half saturation μM	
Myosin	0.14-0.28	10	Dancker (1970) Bremel and Weber (1974)
Troponin C	0.28-0.56	0.2 (2)	Potter et al. (1974), Hartshorne and Pyun (1971)
SR Membranes	0.2	0.3	Fiehn and Migala (1970), Chevalier and Butow (1970), Meissner (1973)

reticulum, the amount would represent a fairly large store. It would allow continuous activation of the muscle provided that recycling occurs sufficiently fast. However, on the basis of radioautographic studies of frog muscles, Winegrad (1968) has estimated the total calcium in the reticulum to be only in the order of 0.6 $\mu mol.ml^{-1}$, which is approximately only 30% of its total calcium content. A calcium store of this size would hardly be sufficient for the activation of the contractile protein if it really needed 0.2 $\mu mol.ml^{-1}$ for full activation (Table 5). However, there is another dilemma which seems difficult to resolve at the moment. If the calcium stored inside the vesicles is present as soluble calcium, an internal concentration between 6 and 20 mM would exist inside the sarcoplasmic membranes. As mentioned above, such high internal calcium concentration severely in-

Table 5. Activating calcium

Requirement µmol/g	Availability µmol/g
0.14-0.28 for 50% activation of SK-muscle	0.6 (Winegrad, 1968) 1.5-2.0

hibits calcium uptake and the energy-yielding hydrolysis of ATP of the isolated vesicular membranes. The calcium-binding protein isolated by MacLennan and Wong (1970), Ostwald and MacLennan (1974) and Ikemoto et al. (1971) from the sarcoplasmic membrane seemed to offer a way out of this difficulty. However, neither the calcium affinity nor the calcium-binding capacity of the calcium-binding protein are high enough to account for a sufficient reduction of the free calcium concentration in the reticulum. Therefore, either we have to assume that the reticulum *in situ* is less sensitive to the inhibiting effect of internal calcium ions, or there exist small molecular weight substances like phosphate that can form complexes with calcium at relatively low free concentrations (Table 6).

Table 6. Calcium in the SR

in vivo	*in vitro*
Accumulated calcium in the SR	Calcium accumulation
$Ca_t \sim 10\text{-}20$ mM	at 1 µM Ca_o^{++} $\quad$ 100-200 nmol $\cdot$ mg^{-1} $\quad\quad Ca_t \quad \sim \quad$ 20- 40 mM
	at 0.03 µM Ca_o^{++} $\quad$ 50- 80 nmol $\cdot$ mg^{-1} $\quad\quad Ca_t \quad \sim \quad$ 10- 16 mM
$Ca_i^{++} \sim 0.1$ mM (Beeler and Reuter, 1970)	Ca_i^{++} upper limit ~ 1 mM 40 nmol $\cdot$ mg^{-1} bound to $\quad\quad\quad\quad\quad\quad\quad$ calcium binding $\quad\quad\quad\quad\quad\quad\quad$ protein
Ca_t $\quad$ concentration of $\quad\quad\quad$ total calcium	$PO_4H_2^{--}$ as calcium binding sub- $\quad\quad\quad\quad\quad\quad\quad$ stance (Weber et al., 1966; Hasselbach and Weber, 1974; MacLennan, 1974; Ikemoto, 1974)

Concerning the proper structural components of the sarcoplasmic reticulum from which calcium ions are released, the cisternal elements are the favourite candidates. This concept is supported by Winegrad's (1970) radioautographic observations. He demonstrated that after a short tetanic stimulation of the muscle the calcium concentration in the region of the cisternae is reduced. The experiments with isolated sarcoplasmic reticulum support this idea rather indirectly by excluding the longitudinal elements of the sarcoplasmic reticulum as calcium-releasing sites. Until now, a calcium release which would fulfil the physiological requirements could not be induced from calcium-filled isolated sarcoplasmic vesicles. The results of some calcium release experiments are given in Fig. 4 and Table 4. Although these experiments are not very conclusive as far as the calcium-releasing structure and mechanisms are concerned, they reveal a number of interesting properties of the calcium-transporting mechanism of the sarcoplasmic mem-

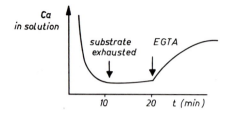

Activation energy 10000 cal/mol ; pH – minimum 6.3
(pH 7.0)

Acceleration: $Cl < Br < NO_3 < ClO_4 < SCN$; Ca Jonophores

Retardation: Succinate ; Glutarate ; Sucrose

Activation : Arsenate ; Activation energy 27000 cal/mol

Jnhibition : Ca_o^{++} (0.2 μM)

Fig. 4. Calcium release from calcium-loaded sarcoplasmic vesicles.
The release-promoting agents are listed

branes. A sudden application of high concentrations of potassium chloride, sodium chloride or sodium succinate produces only moderate effects. A somewhat more pronounced increase of the rate of calcium
release was produced by chaotropic anions. Sudden pH changes are likewise hardly effective. An acceleration of the release was produced by
the application of antibodies against the calcium transport ATPase
in combination with complement (Martonosi et al., 1974). Similar effects were obtained with the calcium ionophore X 537 (Scarpa et al.,
1972). Caffeine, a calcium-releasing substance in the living muscle,
does not release calcium from isolated and purified sarcoplasmic reticulum. Similar negative results were obtained when we tried to imitate
the so-called calcium-stimulated calcium release deduced by Ford and
Podolsky (1972) and Endo et al. (1970) from experiments with striped
fibers. Makinose (1973) found that the addition of calcium to pre-
loaded sarcoplasmic vesicles stimulated a fast calcium exchange but
not a net calcium release. He found that ADP was necessary for this
exchange of calcium. This finding is an important oberservation relevant for the understanding of the mechanism of active calcium translocation. A relatively fast release of calcium was achieved by driving
the pump in the reverse direction, adding EGTA together with inorganic
phosphate and ADP. However, like the other releasing mechanisms the
ADP + Pi-stimulated calcium release cannot account for the fast activation of the contractile system (Temple et al., 1974). Therefore,
we are inclined to assume that we have lost the structures equipped
with the mechanism for calcium release during the isolation and purification of the sarcoplasmic membrane. However, even if we could isolate and purify the releasing structures, it remains an open question
whether these isolated elements were able to store calcium and if the
mechanism by which calcium release is induced would survive the isolation procedure. The extreme sensitivity of these structures is illustrated by the marked morphological change which occurs in the region of the junction between the T-tubules and the cisternal elements
when the environment of the living fiber deviates only slightly from
isotonicity (Birks and Davey, 1972).

III. Calcium Removal

Let us now turn to our last problem, the problem of calcium net removal during relaxation. It must be checked and proved whether the sarcoplasmic reticulum is able to remove the calcium ions required for the activation of the contractile protein during relaxation. To answer this question we have to know the amount of sarcoplasmic membranes in the muscle and the specific rates at which the ATP-driven calcium pump accumulates calcium. For fast-contracting muscles approximately 10 mg sarcoplasmic membranes per ml^{-1} of fiber are considered a good approximation (Peachey, 1965) although the extraction yield does usually not exceed 5 mg.ml^{-1}. Specific uptake rates were measured at physiological calcium concentration in the medium, but in the presence of oxalate or phosphate which allow the pump to run freely in the forward direction (cf. Hasselbach, 1974). Surprisingly these values are in good agreement with those obtained by Inesi and Scarpa (1972) measuring calcium uptake in the absence of calcium precipitating agents at calcium concentrations in the assay media which must be considered higher than those existing in the living muscle (Table 7). From the figures

Table 7. Calcium removal

by Calcium transport

$$2\ Ca_{my} + ATP \longrightarrow 2\ Ca_i + ADP + P$$

Conditions	Rate
Ca_o^{++} 1 µM	
Ca_i^{++} 4 µM	$\mu mol \cdot ml^{-1} \cdot sec^{-1}$
10 mg SR/ml T = 20°C	0.3
Ca_o^{++} 10-100 µM	
Ca_i^{++} 100 µM?	0.4
10 mg SR/ml T = 20°C	

by Calcium binding

$$Ca + M \overset{k_1}{\rightleftharpoons} CaM \quad k_1 = 5 \cdot 10^8 [l \cdot mol^{-1} \cdot sec^{-1}]$$ (Winkler and Eigen, 1972)

Ca_i^{++} 1 µM

[M] 0.2 mM

$v = 5 \cdot 10^8 \cdot 10^{-6} \cdot 2 \cdot 10^{-4}$

$= 0.1\ mol \cdot l^{-1} \cdot sec^{-1}$

$= 100\ \mu mol \cdot ml^{-1} \cdot sec^{-1}$

given in Table 7 it follows that both rates are just marginal to produce a sufficiently fast relaxation. It seems very improbable that muscle activity is regulated by a system with such small capacity in reserve. Due to this uncertainty Ebashi and Endo (1968) proposed that calcium may be removed by an ATP-binding process. Binding can be considered to occur always very fast and the reticulum has a relatively high number of calcium-binding sites. Ebashi's binding concept would require a process which reduces the affinity of the reticulum for calcium during excitation. Such a process seems to be not very probable. Moreover, such an ATP-dependent binding process in the reticulum has been disproved (Fiehn and Migala, 1970). Nevertheless, a fast binding process may be involved in relaxation. Due to the activity of the calcium pump, the external binding sites of the sarcoplasmic membranes remain unoccupied in the resting muscle. Consequently, an amount of calcium corresponding to that by which the contractile protein could have been activated can be bound very quickly. The calculations in Table 7 are based on the rate measurements of Eigen (cf. Winkler and Eigen, 1972) for the formation of calcium-ATP complexes. However, during repetitive contractions the binding capacity of the sarcoplasmic reticulum would be overridden and the slow ATP-dependent calcium translocation step would become the rate-limiting process. Since this process has to replenish the calcium stores in the reticulum from which calcium is released, calcium release and, as a consequence, the activity of the muscle has to adapt to the activity of the calcium pump.

References

Adrian, R.H., Peachey, L.D.: J. Physiol. (Lond.) 235, 103-131 (1973).
Ashley, C.C., Ridgway, E.B.: J. Physiol. (Lond.) 209, 105-130 (1970).
Baker, P.F., Blaustein, M.P., Hodgkin, A.L., Steinhardt, R.A.: J. Physiol. (Lond.) 200, 431-458 (1969).
Beeler, G.W., Reuter, H.: J. Physiol. (Lond.) 207, 191-209 (1970).
Benzanilla, F., Caputo, C., Gonzales-Serratos, H., Venosa, R.A.: J. Physiol. (Lond.) 223, 507-523 (1972).
Birks, R.I., Davey, D.F.: J. Physiol. (Lond.) 222, 95-111 (1972).
Bremel, R.D., Weber, A.: Biochim. Biophys. Acta 376, 366-374 (1975).
Chevalier, J., Butow, R.A.: Biochem. 10, 2733-2757 (1970).
Dancker, P.: Pflügers Arch. 315, 198-211 (1970).
Ebashi, S., Endo, M.: Progr. Biophys. Mol. Biol. 18, 123 (1968).
Endo, M., Tanaka, M., Ogawa, Y.: Nature 228, 34-35 (1970).
Fiehn, W., Migala, A.: Eur. J. Biochem. 20, 245-248 (1970).
Ford, L.E., Podolsky, R.J.: J. Physiol. (Lond.) 223, 21-33 (1972).
Gonzales-Serratos, H.: J. Physiol. (Lond.) 212, 777-799 (1971).
Hagiwara, S., Nakajima, S.: J. Gen. Phys. 49, 793-806 (1966).
Hartshorne, D.J., Pyun, H.Y.: Biochim. Biophys. Acta 229, 698-711 (1971).
Hasselbach, W.: Enzymes, Vol. 10, pp. 432-468. New York: Academic Press 1974.
Hasselbach, W., Makinose, M.: Biochem. Z. 333, 518-528 (1961).
Hasselbach, W., Makinose, M.: Biochem. Z. 339, 94-111 (1963).
Hasselbach, W., Weber, H.H.: Membrane Proteins in Transport and Phosphorylation, pp. 103-111. Amsterdam: North Holland 1974.
Heilbrunn, L.V.: Physiol. Zool. 13, 88-94 (1940).
Hellam, D.C., Podolsky, R.J.: J. Physiol (Lond.) 200, 807-819 (1969).
Hill, A.V.: Proc. Roy. Soc. B 135, 446-453 (1948).
Huxley, A.F.: Ann. New York Acad. Sci. 81, 446-452 (1959).
Ikemoto, N.: J. Biol. Chem. 249, 649-651 (1974).
Ikemoto, N., Bhatnagar, G.M., Gergely, J.: Biochem. Biophys. Res. Commun. 44, 1510-1517 (1971).
Inesi, G., Scarpa, A.: Biochem. 11, 356-359 (1972).
Julian, F.J.: J. Physiol. (Lond.) 218, 117-145 (1971).
Keynes, R.D., Rogas, E., Taylor, R.E., Vergara, J.: J. Physiol. (Lond.) 229, 409-455 (1973).
MacLennan, D.H.: J. Biol. Chem. 249, 980-984 (1974).
MacLennan, D.H., Wong, P.T.S.: Proc. Nat. Acad. Sci. USA 68, 1231-1235 (1971).
Makinose, M.: FEBS Lett. 37, 140-143 (1973).
Makinose, M., Hasselbach, W.: Biochem. Z. 343, 360-382 (1965).
Martonosi, A., Jilka, R., Fortier, F.: Membrane Proteins in Transport and Phosphorylation, pp. 113-124. Amsterdam: North Holland 1974.
Meissner, G.: Biochim. Biophys. Acta 298, 906-926 (1973).

Ostwald, T.J., MacLennan, D.H.: J. Biol. Chem. <u>249</u>, 974-979 (1974).
Peachey, L.D.: J. Cell Biol. <u>25</u>, 209-231 (1965).
Perry, S.V., Cole, H.A., Head, J.F., Wilson, T.J.: Cold Spring Harbor
    Symp. Quant. Biol. <u>37</u>, 251-262 (1972).
Porter, K., Palade, G.E.: J. Biophys. Biochem. Cyt. <u>3</u>, 269-300 (1957).
Portzehl, H.: Verh. Deutsche Gesell. Innere Med., pp. 125-136. München:
    J.F. Bergmann 1965.
Portzehl, H., Caldwell, P.C., Rüegg, C.: Biochim. Biophys. Acta <u>79</u>,
    581-591 (1964).
Portzehl, H., Zaoralek, P., Grieder, A.: Pflügers Arch. ges. Physiol.
    <u>286</u>, 44-56 (1965).
Potter, J., Seidel, J., Leavis, P., Lehrer, S.S., Gergely, J.: Calcium
    Binding Proteins, pp. 129-159. Amsterdam: Elsevier 1974.
Rüdel, R., Taylor, S.R.: J. Physiol. (Lond.) <u>233</u>, 5 (1973).
Scarpa, A., Baldassare, J., Inesi, G.: J. Gen. Physiol. <u>60</u>, 735-749
    (1972).
Schatzmann, H.J.: J. Physiol. (Lond.) <u>235</u>, 551-569 (1973).
Schaub, M.C., Perry, S.V.: Biochem, J. <u>115</u>, 993-1004 (1969).
Taylor, S.R.: The Physiological Bases of Sterling's Law of the Heart.
    Ciba Foundation Symp. 24, 93-116. Amsterdam: North Holland 1974.
Temple, D., Hasselbach, W., Makinose, M.: Naunyn-Schmiedeberg's Arch.
    Pharmakol. <u>282</u>, 187-194 (1974).
Weber, A.: J. Gen. Physiol. <u>57</u>, 50-63 (1971).
Weber, A., Herz, R.: J. Biol. Chem. <u>238</u>, 599-605 (1963).
Weber, A., Herz, R., Reiss, J.: Biochem. Z. <u>345</u>, 329-369 (1966).
Weber, A., Winicur, S.: J. Biol. Chem. <u>236</u>, 3198-3202 (1961).
Winegrad, S.: J. Gen. Physiol. <u>51</u>, 65-83 (1968).
Winegrad, S.: J. Gen. Physiol. <u>55</u>, 77-88 (1970).
Winkler, R., Eigen, M.: Molecular Bioenergetics and Macromolecular
    Biochemistry, pp. 130-148. Berlin-Heidelberg-New York: Springer 1972.

Discussion

Dr. Gergely: What is the basis of your statement that the SR preparations contain mainly the longitudinal component?

Dr. Hasselbach: We usually screen our SR-preparations with the electron
microscope. Cisternal elements, that means elements which are in contact with transverse tubules, are very seldom. The only picture I have
of these structures is five or six years old. I suppose that we lose
the cysternae during purification.

Dr. Weber: Do you think that we have not yet found a major element
functioning in the calcium uptake of living muscle or that reticulum
after isolation no longer functions the same as in living muscle?

Dr. Hasselbach: I feel we are confronted with various difficulties.
The first dilemma concerns the rate of calcium release. In none of our
release experiments we have observed rates as they are required for
excitation. The other dilemma concerns the rate of calcium uptake. The
values deduced from *in vitro* measurements just suffice to account for
relaxation. Perhaps when we isolate the SR we lose some labile calcium
binding material which usually is present inside the vesicles and which
keeps the calcium concentration relatively low. My intention was to
emphasize the many loopholes in our concept of excitation-contraction
coupling.

Dr. Perry: I wonder, Dr. Hasselbach, if you could explain the results
obtained on electrophoresis of whole sarcoplasmic reticulum. Did I

understand you in saying that you could not detect very much calsequestrin? Also were you suggesting that calsequestrin was very similar to serum albumin?

Dr. Hasselbach: We find usually only a very faint band in front of the transport ATPase which corresponds to the calcium precipitating protein (8% of the total protein). It has the mobility of serum albumin. A paper has been published recently in (1974) Biochemistry 13, 3298 by Inesi. He likewise states that calsequestrin moves together with serum albumin.

Dr. Gergely: In our experience calsequestrin always moves in SDS gel electrophoretograms at a rate which differs from that of serum albumin.

Dr. Hasselbach: I do not think that we use an albumin different from yours. - There may be differences due to the buffer systems used for the gel electrophoresis.

Dr. Drabikowski: In our opinion the problem of the number of protein present in SR and being seen in polyacrylamid gels depends greatly on the loading of the gel. When 20 μg of SR protein is put on gel only 100,000 daltons ATPase is found. With 100 μg of protein, besides ATPase, at least 3 other proteins are detected. The biggest in amount is the protein with MW 45,000. The other protein moving at the level of serum albumin is present in much smaller amounts (see Fig.)

Dr. Hasselbach: There is no doubt that the main band in front of the calcium transport protein moves in Tris-bicine buffer pH 8.2; 0.1% SDS together with serum albumin.

Protein pattern of SR membrane

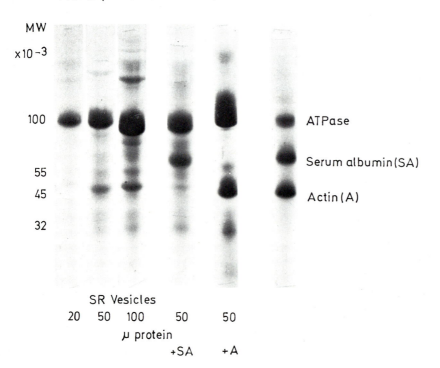

The Role of the Interaction of Ca^{2+} with Troponin in the Regulation of Muscle Contraction

J. D. Potter, B. Nagy, J. H. Collins, J. C. Seidel, P. Leavis,
S. S. Lehrer, and J. Gergely

A. Introduction

In introducing this paper, that is dedicated to the memory of Professor
H.H. Weber and deals with some aspects of the interaction of calcium
and one of the regulatory proteins of muscle, it is fitting to recall
that the Institute of which Professor Weber was Director pioneered in
the development of our current concepts concerning the regulation of
muscle contraction by Ca^{2+}. This is attested to in greater detail by
Professor Hasselbach's paper at this Symposium; it is also a pleasant
duty to recall that it was the work of Annemarie Weber, independently
carried out in the United States, that played a key role in bringing
to light the important function of the free Ca^{2+} concentration in the
regulation of actin-myosin interaction. Her work, and that of Ebashi
and his colleagues (for reviews see Ebashi and Endo, 1968; Ebashi et
al., 1969; Weber and Murray, 1973) led to the clear recognition that
the interaction of actin and myosin, the key step in the contraction
process, depends on tiny amounts of free Ca^{++}. Ebashi and his colleagues
went on to show that while the system consisting of purified actin and
myosin had no calcium sensitivity, in the presence of what at first
was known as native tropomyosin the interaction between actin and myosin
- reflected in superprecipitation and actin activation of myosin ATPase
in the presence of Mg^{2+} - is dependent on Ca^{++}. Native tropomyosin was
soon recognized to consist of Bailey's tropomyosin (1948) and a new
entity for which the name troponin (Tn) was coined (Ebashi and Kodama,
1965; Ebashi et al., 1972). Further work established that troponin is
complex of three subunits (Greaser and Gergely, 1971), now generally
referred to as troponin C (TnC) ($M_r \sim 18,000$), troponin I (TnI) ($M_r \sim$
21,000) and troponin T (TnT) ($M_r \sim 36,000$) (Greaser and Gergely, 1973;
Greaser et al., 1972). TnI is identical with the inhibitory component
previously studied by Perry and his colleagues (Perry et al., 1972;
Wilkinson et al., 1972). TnT is characterized by its ability to form
a strong complex with tropomyosin (Cohen et al., 1972; Greaser and
Gergely, 1973; Greaser et al., 1972), and TnC is the component to which
the Ca-combining ability of troponin is attributable (Greaser and Gergely, 1971, 1973; Hartshorne and Pyun, 1971; Potter et al., 1974).

As discussed in H.E. Huxley's paper, in this volume, tropomyosin and
troponin are localized in the thin filaments, and the original suggestion by Ebashi and his colleagues (1969) that one tropomyosin molecule
extends over a length of seven actin units and that each tropomyosin
is complexed with one troponin has been verified through the analysis
of the stoichiometry of the components in myofibrils by means of gel
electrophoresis (Potter, 1974). These components are present in thin
filaments of vertebrate muscles; but troponin is absent from muscles
of a number of species including molluscs, in which the interaction
of actin and myosin is regulated by the combination of Ca^{2+} with myosin

itself (Kendrick-Jones, 1974; Lehman et al., 1972; Regenstein and
Szent-Györgyi, 1975) (see also Kendrick-Jones' presentation in this
Symposium). There are indications that Ca^{2+} may also exert a direct
effect on the myosin cross-bridges of vertebrate muscle, and there are
a number of investigations between myosin and calcium in the same concentration range as that required for *in vitro* activation (Balint et
al., 1975; Bremel and Weber, 1975; Gaffin and Oplatka, 1974; Morimoto
and Harrington, 1974; Potter, 1975a; Weber and Oplatka, 1974).

The generally held view concerning the mechanism of the regulation of
actin and myosin interaction is based on the finding that tropomyosin
can change its position within the thin filaments (Haselgrove, 1972,
1975; Huxley, 1972; Parry and Squire, 1973; Spudich et al., 1972; Vibert
et al., 1972; Wakabayashi et al., 1975). In relaxed muscle $[Ca^{2+}]$ is
kept at about 0.1 μM (see Hasselbach in this volume) by the Ca^{2+}-pump
of sarcoplasmic reticulum. Excitation of the muscles leads to the release of Ca^{2+} from the sarcoplasmic reticulum and to a rise in the
free Ca^{2+} concentration. Ca^{2+} then binds to troponin, and induces the
movement of troponin from the blocking to the active position. The
structural evidence for this movement has been discussed by H.E. Huxley
at this Symposium. The purpose of our paper is to deal in some detail
with the question of the interaction of calcium with troponin, particularly the interaction of Ca^{2+} with the Ca^{2+}-binding component of troponin. An understanding of the details of the combination of Ca^{2+} with
troponin and of some of the changes that occur in TnC on the interaction will, one hopes, throw light on the molecular mechanism of the
regulatory process.

B. Ca^{2+} and Mg^{2+} Binding to Troponin

I. Equilibrium Dialysis

Recently (Potter and Gergely, 1975; Potter et al., 1974) we have carried out equilibrium dialysis studies on both TnC and unfractionated
Tn, the Ca^{2+} concentration being controlled by ethylene glycol bis
(β-aminoethyl ester) (EGTA). The number of binding sites and binding
constants was evaluated by fitting the data with a least-squares computer technique without recourse to the transformation required for
Scatchard plots. We sought to obtain the best fit with the smallest
number of classes of independent binding sites. The agreement between
data and calculation is illustrated in Fig. 1. The constants derived
from these studies are shown in Table 1. Two classes of binding sites
exist with two sites in each class, giving a total of 4 sites for TnC.
If one uses the experimental stoichiometry of 1:1:1 for TnC:TnI:TnT
(Potter, 1974), the total binding to Tn can be accounted for by the
binding to TnC. The affinity of binding to TnC is increased by combination with TnI, and it appears that this interaction also determines
the affinity for Ca^{2+} in the Tn complex. Earlier work on Ca^{2+} binding
was hampered by the lack of highly purified preparations, and technical
problems arose both because of the difficulty in controlling Ca^{2+} at
the low levels necessary for determining the properties of high affinity binding sites (Drabikowski and Barylko, 1971; Ebashi et al.,
1968; Fuchs, 1972; Fuchs and Briggs, 1968; Hartshorne and Pyun, 1971)
and because of difficulties inherent in the evaluation of the binding
data by means of Scatchard plots (Klotz, 1974). Previous binding studies carried out in the presence of Mg^{2+} were also complicated by un-
recognized effects of Mg^{2+} on Ca^{2+} binding. In the presence of 2 mM
Mg^{2+} the affinity of the class with a higher affinity is reduced and
one obtains the best fit with a single class containing four binding

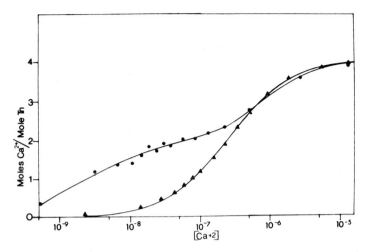

Fig. 1. Ca^{2+} binding to Tn and the effect of Mg^{2+} on it. Equilibrium dialysis, [KCl] = 150 nM. Key: ●, no Mg^{2+}; ▲, 2 mM Mg^{2+}. The solid curve represents the best fit of the data with the computerized least -squares method. In the absence of Mg^{2+} the best fit was obtained with $n_1 = 2.05$, $K_1 = 3.7 \times 10^8$ M^{-1} and $n_2 = 2.37$, $K_2 = 1.08 \times 10^6$ M^{-1}. In the presence of Mg^{2+} the best fit was obtained with n = 4.3, K = 4.03 $\times 10^6$ M^{-1}, representing a single class of four independent equivalent sites. For details see Potter and Gergely (1975). (Reproduced with permission)

Table 1. Ca^{2+} binding

	[MgCl$_2$]	n_1 mol/mol	$10^{-6} \times K_1$ M^{-1}	n_2 mol/mol	$10^{-6} \times K_2$ M^{-1}
TnC	none(4)	1.8±0.05	21 ±3.5	2.1±0.1	0.32±0.06
TnC	2 mM(7)	1.8±0.04	2.8±0.6	2.1±0.04	0.11±0.02
TnC-TnI	none(3)	1.9±0.06	220 ±35	2.3±0.1	3.5 ±0.6
TnC-TnI	2 mM(3)	4.3±0.06	4.3±0.9		
Tn	none(4)	2.1±0.15	530 ±75	2.0±0.1	4.9 ±0.35
Tn	2 mM(4)	4.1±0.25	5.4±0.45		

Equilibrium dialysis experiments: the KCl concentration was 100 mM for TnC and TnC-TnI and 150 mM for Tn. TnC and TnI were mixed in a 1:1 molar ratio. The values of n and K are averages with standard error of the mean, determined by the least-squares procedure, of the individual values of n and K from several experiments. The number of determina- tions is shown in parentheses. The binding to Tn is calculated assuming that Tn is a 1:1:1 complex of TnC, TnI and TnT (Potter, 1974). (Repro- ducted with permission from Potter and Gergely, 1975)

sites. The data can be interpreted in terms of competition between Ca^{2+} and Mg^{2+} at the high-affinity binding sites. There is good agreement between the Mg^{2+} binding constant determined from direct binding stud- ies and that calculated from the Ca^{2+} binding data assuming competition: both values are about 5×10^3 M^{-1}. Direct Mg^{2+} binding studies show that there are four binding sites for Mg^{2+}, two of which can be identified with the Ca^{2+} binding sites of higher affinity.

Thus there are a total of six divalent metal binding sites on TnC. We refer to them as the Ca^{2+}-Mg^{2+} sites, the Ca^{2+}-specific sites and the Mg^{2+}-specific sites. We have chosen the simplest way of describing the binding data, i.e. in terms of two sets of independent binding sites. One could probably invoke "negative cooperativity" to account for the two binding sites of lower affinity, but the experimental data would not permit us to distinguish between the two hypotheses. In what follows we shall discuss those sites that bind Ca^{2+}, since nothing can be said at this stage about the role, if any, of the Mg^{2+}-specific sites.

II. Structural Aspects

The location of the four calcium binding sites in the amino acid sequence has been deduced (Collins et al., 1973) by comparing the amino acid sequence of TnC with the sequence and three-dimensional structure (Kretsinger and Nockolds, 1973) of a homologous carp calcium binding protein (CBP) commonly known as parvalbumin. CBP contains two similar calcium-binding regions related to each other by a two-fold axis of symmetry. Each site lies in a loop between two short α-helical segments (Fig. 2). Collins et al. (1973) suggested that the four calcium-binding regions in TnC are similar to those of CPB (Fig. 3), which conclusion has subsequently been confirmed by others (Barker and Dayhoff, 1975; Kretsinger, 1975; Pechere et al., 1973; Tufti and Kretsinger, 1975; Weeds and McLachlan, 1974). As suggested by Kretsinger and Barry (1975)

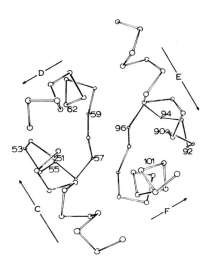

Fig. 2. Three-dimensional representation of Ca^{2+} binding sites in carp muscle CBP. Helices C and D flank one calcium binding loop; helices E and F flank the other. The CD region is related to the EF region by an approximate 2-fold axis. Kretsinger and his colleagues refer to the domain including pairs of helices such as E and F and the loop in between as "hands". (Reproduced with permission from Kretsinger and Nockolds, 1973)

regions I and II, on the one hand, and III and IV, on the other, in TnC correspond to the CD plus EF structure shown in Fig. 2. The two-fold axes would be coincident but antiparallel and the I-II pair would be rotated with respect to the III-IV pair by about 90°. As noted by Kretsinger (1974), the two sites in the N-terminal half of TnC contain glycine residues (which have no side chain); this could cause the calciums to be more exposed to solvent than those in the C-terminal half of TnC. One would then expect that the N-terminal half of the molecule contains Ca^{2+} bound with a lower affinity, while the C-terminal half contains two very tightly bound calciums. However, Weeds and McLachlan

```
CBP  (50- 63)   CD: ASP-gln-ASP-lys-ser-GLY-phe-ILE-glu-glu-asp-GLU
     (89-101)   EF: ASP-ser-ASP-gly-asp GLY-lys-ILE-gly-val-asp-GLU

TnC  (27- 38)    I: ASP-ala-ASP-gly-gly-GLY-asp-ILE-ser-val-lys-GLU
     (63- 74)   II: ASP-glu-ASP-gly-ser-GLY-thr-ILE-asp phe-glu-GLU
    (103-114)  III: ASP-arg-asn-ala-asp-GLY tyr-ILE asp-ala glu-GLU
    (139-150)   IV: ASP-lys-asn-asn-asp-GLY arg-ILE-asp-phe-asp-GLU
```

Fig. 3. Sequences of rabbit TnC and carp muscle CBP in the segments corresponding to the calcium-binding sites of CBP. Residues (in CD and EF) involved in calcium-binding in CBP are underlined. Residues identical in five or more regions are capitalized. Each calcium binding region in TnC is flanked by a pair of regions that appear α-helical by homology with CBP. The two α helices adjacent to region III in TnC contain 3 phe residues and the single cysteine (residue 98) of TnC. (Sequences taken from Collins et al., 1973)

(1974) have suggested that the calcium binding sites in the N-terminal half could bind Ca^{2+} more tightly because they would be more flexible and unconstrained than the C terminal sites. This question may be resolved if the recently described (Mercola et al., 1975) crystals of TnC prove to be suitable for the determination of the three-dimensional structure by X-ray diffraction. The recent finding (Collins, 1974) that myosin alkali light chains (which do not bind calcium) are also homologous with TnC raises many interesting questions on the relationship of actin-linked and myosin-linked regulatory systems. It is now generally agreed (Barker and Dayhoff, 1975; Collins, 1974; Pechere et al., 1973; Tufti and Kretsinger, 1975) that TnC, CBP and alkali light chains all evolved from a common ancestor. This common ancestor was probably a TnC-like protein which arose through duplication and reduplication of a small ($M_r \sim 4,000$) ancestral calcium binding protein (Collins, 1974; Pechere et al., 1973; Weeds and McLachlan, 1974). Myosin also contains another small subunit, the so-called DTNB light chain, which can replace the calcium-sensitizing light chain of molluscan myosin (Kendrick-Jones, 1974). Current sequence studies (Collins, 1975) on the DTNB light chain show that it too is homologous with TnC, and a single calcium binding site has been tentatively located at residue 37-48 in the sequence (Collins, unpublished).

C. Conformational Changes

Indications of changes in the structure of TnC occurring upon interaction with metals have been obtained with a variety of techniques (Ebashi et al., 1974; Gruda et al., 1973; Head and Perry, 1974; Kawasaki and Van Eerd, 1972; Murray and Kay, 1972; Potter et al., 1974; Van Eerd and Kawasaki, 1972, 1973). Measurements of circular dichroism show that removal of Ca^{2+} or Mg^{2+} from the medium by means of chelation reduces the α-helix content by about 50%. The fact that the large change in α-helix content can be obtained upon adding either Ca^{2+} or Mg^{2+} is an indication of the non-specific character of this change. Consistent with the binding data, the change occurs in the range of Ca^{2+} concentration (Kawasaki and Van Eerd, 1972; Van Eerd and Kawasaki, 1972) characteristic of the Ca^{2+}-Mg^{2+} binding sites.

TnC contains only one reactive sulfhydryl group (Cys-98) which is in the α-helical portion adjacent to one of the putative Ca^{2+} binding sites. The reactivity of this sulfhydryl group is greatly reduced in the presence of Ca^{2+} or Mg^{2+}, as was indicated by the difficulty in putting a thiol-specific spin label on TnC unless the divalent cations have been chelated. The mobility of our iodoacetamide spin label attached ty Cys-98 is reduced on adding either Ca^{2+} or Mg^{2+} (Potter et al., 1973, 1974). Ca^{2+} alone produces a slightly larger change than does Mg^{2+} and addition of Ca^{2+} in the presence of Mg^{2+} produces an additional change. Again it appears that a conformational change underlying the mobility change of the attached label is caused by a combination of Ca^{2+} or Mg^{2+} at high affinity sites.

As recently shown (Potter, 1975b) the reactivity of the sulfhydryl group measured by its reaction with DTNB changes with $[Ca^{2+}]$ in the same way as does the mobility of the spin label attached to it (Fig. 4). These facts suggest that one of the regions that undergoes changes when Ca^{2+} binds in the high affinity range is that containing the Ca^{2+} binding site (Fig. 3) adjacent to Cys-98.

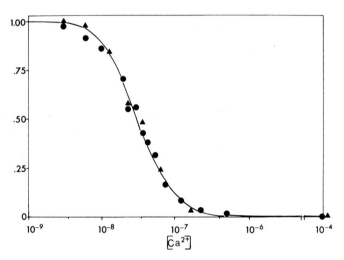

Fig. 4. Changes in the electron spin resonance spectrum of spin-labelled TnC and the reactivity of the Cys-98 -SH group as a function of pCa. Conditions: 100 mM KCl, 10 mM imidazone, pH 7.0, and 2 mM EDTA. Enough $CaCl_2$ was added to produce the Ca^{2+} concentration shown on the abscissa. Ordinate: relative values. ▲ , reaction rate with DTNB; ● , mobility of spin label deduced from a change in the height of the upfield peak of the esr spectrum. For details of spin-labelling and esr measurements see Potter et al. (1974). The reaction rate with DTNB was measured by following the color development at 412 nm; TnC, 1 mg/ml; DTNB, 0.2 mM. Rates were evaluated from first order semilogarithmic plots. Temperature, 25°

The sensitivity of this region to conformational changes is also revealed by the increase in the fluorescence af a dansyl-S-mercuric cysteine label (Leavis and Lehrer, 1974) attached to the same Cys-98 (Leavis et al., 1974); the spectral changes attributable to Tyr (Kawasaki and Van Eerd, 1972; Van Eerd and Kawasaki, 1972) may in part be due to changes in the environment of Tyr-109, also located in region III. The Ca^{2+} concentration range in which the change occurs again corresponds to the Ca^{2+}-Mg^{2+} binding sites. However, in this case the conformational change presumably underlying the change in fluorescence seems to be specific for Ca^{2+}, since no enhancement occurs in the presence of Mg^{2+} alone. Nevertheless, the concentration range in which Ca^{2+} causes fluorescence enhancement is shifted to higher values in the presence of Mg^{2+}, which is consistent with the participation of the $Ca^{2+}-Mg^{2+}$ sites. The fluorescent label is also sensitive to changes produced by Ca^{2+} binding to the lower affinity site, as shown by the biphasic character of the fluorescence response, corresponding to the two Ca-binding constants. In the presence of TnI the response to Ca^{2+} of TnC carrying the fluorescent label occurs at a lower Ca^{2+} concentration, but in this case the difference between Ca^{2+} and Mg^{2+} disappears. This is in contrast to the behavior of the spin label attached to TnC (Potter et al., 1974) which, even when complexed with TnI, shows a decrease in mobility on adding Ca^{2+} in the presence of Mg^{2+}. In attempting to interpret conformational changes observed on adding Ca^{2+} to TnC or Tn from a functional point of view, it should be borne in mind that the biological activity of Ca^{2+} in the tropomyosin-troponin actin-myosin system is observed in the presence of Mg^{2+} and that in "switching on" the actin myosin system Mg^{2+} cannot substitute for Ca^{2+}. Therefore, conformational changes likely to be involved in the biological mechanism are those that are produced specifically by Ca^{2+} in the presence of Mg^{2+}. Conformational changes that can be produced by either Ca^{2+} or Mg^{2+} may be needed for establishing the proper three-dimensional structure of TnC but are not likely to be directly involved in the Ca^{2+} specific switching mechanism.

Further insight into the mechanism by which calcium binding to the troponin produces conformational changes, particularly in terms of the identification of certain calcium-binding sites, can be obtained from studies of Ca^{2+} induced changes in TnC in urea (Nagy et al., 1975), as well as from some studies of ours (unpublished) on a cyanogen bromide fragment (Collins et al., 1973) of troponin C (CB-9, res. 84-135) containing the putative calcium-binding site in region III and the two adjacent α-helical segments. This fragment contains the Cys-98, Tyr-109, as well as 3 Phe residues.

In 6 M urea the α-helix content of TnC, as indicated by the ellipticity at 222 nm, is lost (Table 2), but on addition of Ca^{2+} about half of the

Table 2. Effect of calcium on the circular dichroism of TnC with and without urea

	+Ca	-Ca	Δ
No urea	- 14,000	- 6,500	7,500
6 M urea	- 8,000	- 1,400	6,500

Values are given as mean residue ellipticity $[\theta]_{222 \text{ nm}}$, deg cm^2 dmol^{-1}

The solution contained 0.1 M KCl, 2 mM HEPES, pH 7.0 and 1 mg of TnC per ml (0.05 mM). Calcium, when added, 0.1 mM. In experiments with no Ca^{2+}, 1 mM EDTA was present.

original helix content returns. This change is essentially the same as that observed upon adding Ca^{2+} to Ca^{2+}-free TnC in the absence of urea, although the Ca^{2+}-free TnC in the absence of urea still has about 50% of the helix content of Ca^{2+}-TnC. This suggests that Ca^{2+} binding to those sites that cause the structural change is unaffected by the disorganization of the remaining α-helical regions and that the α-helical change taking place on calcium binding in native TnC is also likely to be localized near the Ca^{2+} binding site(s). In the native state essentially all the α-helical change, as well as the changes discussed above, including the decrease in reactivity of sulfhydryl groups, changes in spin-labelled mobility, and changes in the spectra of tyrosine and pheylalanine occur on binding of Ca^{2+} to the high affinity Ca^{2+}-Mg^{2+} sites. In view of the fact that the same reactivity and spectral changes occur also in urea, one might conclude that one of the high affinity sites is located in the loop adjacent to Cys-98. Upon adding Ca^{2+}, the 52-residue fragment (CB-9) containing Cys-98 and one Tyr also shows changes in helicity, in the spectral properties attributable to alterations in the environment of both tyrosine and phenylalanine residues, in the mobility of the spin label attached to Cys-98, and in the reactivity of the -SH group. The fact that all the responses to Ca^{2+} characteristic of binding Ca^{2+} to the Ca^{2+}-Mg^{2+} sites in intact TnC are observed with the CB-9 fragment supports the identification of the site adjacent to Cys-98 as one of the high affinity sites. However, the affinity of Ca^{2+} to the fragment appears to be weaker than to native TnC. This, one could argue, is an indication that the fragment contains what originally would be the low affinity Ca-specific sites. The more likely possibility is that the original high Ca^{2+} affinity of the intact molecule is partly attributable to interaction of the site with other portions of the molecule, which would be absent in the fragment.

D. ATPase Activity

To gain further insight into the role of various Ca^{2+} binding sites in the control of the actin myosin interaction, we reinvestigated Ca^{2+} dependence of the myofibrillar ATPase activity. On the basis of such studies Bremel and Weber (1972) have previously concluded that Ca^{2+} binding to lower affinity binding sites may be crucial in the activation process. With the use of Ca^{2+} binding constant for troponin, we calculated what the Ca^{2+} dependence of the myofibrillar ATPase activity would be, assuming that binding constants for troponin are applicable to the complete myofibrillar system, depending on whether the activation of the actomyosin ATPase requires Ca^{2+} binding to all Ca^{2+} sites, to the Ca^{2+}-Mg^{2+} sites alone, or to the Ca^{2+}-specific sites alone, at two free Mg^{2+} concentrations. As seen from Fig. 5, on changing the Mg^{2+} concentration from 30 μM to 2 mM (A → B) a considerable shift should occur in the range of activating $[Ca^{2+}]$ if only the high affinity sites were involved; and a significant but smaller shift (B → C) if binding to all four sites were required. No shift is expected if only the low affinity Ca^{2+}-specific sites were involved. Experimental data show that changing the Mg^{2+} concentration by almost a factor of 100 did not produce a change in dependence of myofibrillar ATPase on $[Ca^{2+}]$ (Fig. 6). This suggests that the switching on of the ATPase activity requires the binding of Ca^{2+} only to the Ca-specific sites. A more detailed analysis also shows that the character of the binding curve is consistent with both low affinity Ca-specific sites being necessary for activation. The Ca^{2+}-dependence of ATPase activity shows a much steeper slope at midpoint than would be expected from independent non-cooperative binding. Solaro et al. (1974), using cardiac myofibrils, observed a $[Ca^{2+}]$-dependence of myofibrillar ATPase activity similar to that described here. The apparent cooperativity in the effect of Ca^{2+} on ATPase would be the result of the requirement that two Ca^{2+}-binding sites be occupied on troponin for activity. Our results agree with those of Bremel and Weber (1972) to the extent that the involvement of sites of lower affinity is established, but in contrast to their conclusions it appears to us that the activation of the actomyosin system does not require binding of Ca^{2+} to the high

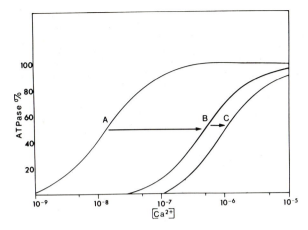

Fig. 5. Calculated dependence of myofibrillar ATPase on the concentration of free Ca^{2+} and Mg^{2+}. Curves calculated for the two different
$[Mg^{2+}]$ (low $[Mg^{2+}]$ = 31.5 µM, high $[Mg^{2+}]$ = 2.15 mM) used in the experiment depicted in Fig. 6. A, regulation requiring Ca^{2+} binding to
both of the Ca^{2+}-Mg^{2+} sites, low $[Mg^{2+}]$; B, regulation requiring Ca^{2+}
binding to both Ca^{2+} - Mg^{2+} sites, high $[Mg^{2+}]$; the same curve applies
to the case of regulation requiring Ca^{2+} binding to all four sites,
low $[Mg^{2+}]$, or regulation requiring Ca^{2+} binding to both Ca^{2+}-specific
sites, high and low $[Mg^{2+}]$; C, Ca^{2+} required at 4 sites, high $[Mg^{2+}]$.
Arrows show shift on raising $[Mg^{2+}]$ from 3..5 µM to 2.15 mM. (Reproduced with permission from Potter and Gergely, 1975)

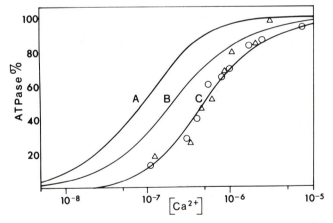

Fig. 6. Dependence of myofibrillar ATPase activity on concentration of
free Ca^{2+} and Mg^{2+}. ATPase activity determined at 25° in a medium containing 150 mM imidazole, pH 7.0, 4 mM EGTA and 5 mM ATP. In the experiments with $[Mg^{2+}]$ = 31.5 µM, 30 mM KCl and 1 mM $MgCl_2$ were present; at
$[Mg^{2+}]$ = 2.15 mM, 9 mM KCl and 7 mM $MgCl_2$ were added. The [KCl] was varied to keep ionic strength constant. $CaCl_2$ was added to achieve the free
Ca^{2+} concentration indicated on the abscissa (for details see Potter and
Gergely, 1975). Experimental data: O, 31.5 µM $[Mg^{2+}]$; Δ, 2.15 mM $[Mg^{2+}]$.
Curve A, calculated Ca^{2+} dependence expected for regulation by Ca^{2+}
binding to either of the Ca^{2+}-specific sites; B, the expected dependence
for binding to one particular Ca^{2+} specific site; C, expected dependence
for Ca^{2+} binding to both of the Ca^{2+}-specific sites. (Reproduced with
permission from Potter and Gergely, 1975)

affinity sites. At present it is somewhat puzzling that studies dealing with the Ca^{2+} dependence of tension development in various muscle fiber systems (Hellam and Podolsky, 1969; Julian, 1971) and some ATPase studies (Weber and Herz, 1963) have shown much steeper changes on changing the Ca^{2+} concentration. In contrast to our results on ATPase activity, as the free $[Mg^{2+}]$ is increased the $[Ca^{2+}]$ required for tension development is also increased (Ebashi and Endo, 1968; Kerrick and Donaldson, 1972). It would be of considerable interest to restudy this question under conditions identical with those used for our ATPase measurements.

E. Mechanism of Activation

New light has been thrown on the mechanism by which the conformational changes induced by Ca^{2+} in TnC finally lead to the movement of tropomyosin and activation of the actin-myosin interaction by studies on the effect of Ca^{2+} on binding of troponin subunits to each other and to tropomyosin and actin (Hitchcock, 1975; Hitchcock et al., 1973; Margossian and Cohen, 1973; Potter and Gergely, 1974; Regenstein and Szent-Györgyi, 1975). The results can be briefly recapitulated as follows. The only interaction that shows Ca^{2+} dependence is the binding of the complex of TnC and TnI to the actin-tropomyosin complex. Since TnT does combine strongly with tropomyosin and since TnI can combine with actin and tropomyosin, it seems most likely that in the absence of Ca^{2+} the TnI complex is attached both to tropomyosin via its TnT component and to actin, actually to a site made up jointly of actin and tropomyosin, via its TnI component. The change produced then in TnC upon combination with Ca^{2+} would lead to a detachment of the TnI from the site made up of tropomyosin and actin.

At this point it will be useful to recall that X-ray evidence and the three-dimensional reconstruction of electron micrographs indicate that the position of the tropomyosin molecules along the actin filament changes depending on whether a muscle is activated or relaxed (Haselgrove, 1975; Parry and Squire, 1973; Spudich et al., 1972; Vibert et al., 1972). In relaxed muscle the tropomyosin would be in a position that would block the combination of the subfragment 1 portion of myosin with actin, whereas in activated muscle, the tropomyosin would move toward the axis of the actin filaments, thereby making it possible for the myosin head, bearing the ADP-P_i complex resulting from the hydrolysis of ATP (see Trentham's paper in this volume), to interact with actin. The implications of this arrangement, whereby the tropomyosin molecule controls seven actin monomers, have been subjected to detailed analysis by A. Weber and her colleagues (1972), in particular from the point of view of the cooperative aspects of actin-myosin interaction.

It seems that the role of troponin in resting muscle is to anchor tropomyosin in the relaxed position. On activation Ca^{2+} is released from the sarcoplasmic reticulum, combines with TnC, and induces a change in it, eventually causing the detachment of the TnI moiety, which makes it possible for tropomyosin to move. A comparison of the state of affairs in skeletal muscle with those in molluscan muscle (Parry and Squire, 1973) suggests that the movement of tropomyosin is not due to the fact that the position assumed by tropomyosin in the active state is thermodynamically favored and the position in the relaxed state is due to the restraining effect of troponin in the absence of Ca^{2+}. For in relaxed molluscan muscles too, which do not contain troponin (Lehman et al., 1972), tropomyosin is found in approximately the same position as in relaxed striated muscle of higher organisms (Parry and

Squire, 1973). Thus, one would have to postulate a conformational change induced in tropomyosin, or perhaps in actin, by the combination of Ca^{2+} with TnC that would initiate the movement of tropomyosin. On the other hand, the recent work of Wakabayashi et al. (1975) on rabbit proteins indicates that tropomyosin would not be in the blocking position in reconstituted filaments containing actin and tropomyosin but no Tn. On the basis of these data release of the TnI-anchor from the tropomyosin-actin site would be sufficient to allow the movement of tropomyosin to take place without conformational changes in tropomyosin.

Acknowledgments. This work was supported by grants from the NIH (HL-5949, HL-15391, AM-11677, HL-17464); the National Science Foundation (GB 43484, GB 24316; GB 38380) the American Heart Association; the Muscular Dystrophy Associations of America, Inc. and the American Heart Association, Massachusetts Affiliate. J.P. is an Established Investigator of the American Heart Association. P.L. is a Postdoctoral Research Fellow of the Muscular Dystrophy Associations of America, Inc.

References

Bailey, K.: Biochem. J. 43, 271-279 (1948).
Balint, M., Sreter, F.A., Wolf, I., Nagy, B., Gergely, J.: J. Biol. Chem. 250, 6168-6177 (1975).
Barker, W.C., Dayhoff, M.O.: Biophys. J. 15, 121a (1975).
Bremel, R.D., Murray, J.M., Weber, A.: Cold Spring Harbor Symp. Quant. Biol. 37, 267-275 (1972).
Bremel, R.D., Weber, A.: Nature (New Biol.) 238, 97-101 (1972).
Bremel, R.D., Weber, A.: Biochim. Biophys. Acta 376, 366-374 (1975).
Cohen, C., Caspar, D.L.D., Johnson, J.P., Nauss, K., Margossian, S.S., Parry, D.A.D.: Cold Spring Harbor Symp. Quant. Biol. 37, 287-297 (1972).
Collins, J.H.: Biochem. Biophys. Res. Commun. 58, 301-308 (1974).
Collins, J.H.: Fed. Proc. 34, 539 (1975).
Collins, J.H., Potter, J.D., Horn, M.J., Wilshire, G., Jackman, N.: FEBS Lett. 36, 268-272 (1973).
Drabikowski, W., Barylko, B.: Acta Biochem. Polonica 18, 353-366 (1971).
Ebashi, S., Endo, M.: Progr. Biophys. Molec. Biol. 18, 123-183 (1968).
Ebashi, S., Endo, M., Ohtsuki, I.: Quart. Rev. Biophysics 2, 351-384 (1969).
Ebashi, S., Kodama, A.: J. Biochem. 58, 107-108 (1965).
Ebashi, S., Kodama, A., Ebashi, F.: J. Biochem. 58, 465-477 (1968).
Ebashi, S., Ohnishi, S., Abe, S., Maruyama, K.: J. Biochem. 75, 211-213 (1974).
Ebashi, S., Ohtsuki, I., Mihashi, K.: Cold Spring Harbor Symp. Quant. Biol. 37, 215-223 (1972).
Ellman, G.L.: Arch. Biochem. Biophys. 82, 70-77 (1959).
Fuchs, F.: Int. J. Peptide Res. 4, 147-149 (1972).
Fuchs, F., Briggs, F.N.: J. Gen. Physiol. 51, 655-676 (1968).
Gaffin, L., Oplatka, A.: J. Biochem. 75, 277-281 (1974).
Greaser, M.L., Gergely, J.: J. Biol. Chem. 246, 4226-4233 (1971).
Greaser, M.L., Gergely, J.: J. Biol. Chem. 248, 2125-2133 (1973).
Greaser, M.L., Yamaguchi, M., Brekke, C., Potter, J., Gergely, J.: Cold Spring Harbor Symp. Quant. Biol. 37, 235-244 (1972).
Gruda, J., Therien, H.M., Lermakian, E.: Biochem. Biophys. Res. Commun. 52, 1307-1313 (1973).
Hartshorne, D.J., Pyun, H.Y.: Biochim. Biophys. Acta 229, 698-711 (1971).
Haselgrove, J.C.: Cold Spring Harbor Symp. Quant. Biol. 37, 341-352 (1972).

104

Haselgrove, J.C.: J. Molec. Biol. 92, 113-143 (1975).
Head, J.F., Perry, S.V.: Biochem. J. 137, 145-154 (1974).
Hellam, D.C., Podolsky, R.J.: J. Physiol. 200, 807-819 (1969).
Hitchcock, S.E.: Eur. J. Biochem. 52, 255-263 (1975).
Hitchcock, S.E., Huxley, H.E., Szent-Györgyi, A.G.: J. Molec. Biol. 80, 825-836 (1973).
Huxley, H.E.: Cold Spring Harbor Symp. Quant. Biol. 37, 361-376 (1972).
Julian, F.J.: J. Physiol. 218, 117-145 (1971).
Kawasaki, Y., Van Eerd, J.P.: Biochem. Biophys. Res. Commun. 49, 898 to 905 (1972).
Kendrick-Jones, J.: Nature 249, 631-634 (1974).
Kerrick, W., Donaldson, S.: Biochim. Biophys. Acta 275, 117-122 (1972).
Klotz, I.M.: Accounts of Chemical Res. 7, 162-168 (1974).
Kretsinger, R.H.: In: Perspectives in Membrane Biology (eds. S. Estra-da-O, C. Gitler), pp. 229-262. New York: Academic Press 1974.
Kretsinger, R.H., Barry, C.D.: Biochim. Biophys. Acta. In press (1975).
Kretsinger, R.H., Nockolds, C.E.: J. Biol. Chem. 248, 3313-3326 (1973).
Leavis, P., Lehrer, S.S.: Biochemistry 13, 3042-3048 (1974).
Leavis, P., Lehrer, S.S., Potter, J.D., Gergely, J.: Fed. Proc. 33, 1293 (1974).
Lehman, W., Kendrick-Jones, J., Szent-Györgyi, A.G.: Cold Spring Harbor Symp. Quant. Biol. 37, 319-330 (1972).
Margossian, S.S., Cohen, C.: J. Molec. Biol. 81, 409-413 (1973).
Mercola, D., Bullard, B., Priest, J.: Nature 254, 634-635 (1975).
Morimoto, K., Harrington, W.: J. Molec. Biol. 88, 693-709 (1974).
Murray, A.C., Kay, C.M.: Biochemistry 11, 1622-2627 (1972).
Nagy, B., Potter, J.D., Gergely, J.: Abstract, Biophys. Soc., p. 35a (1975).
Parry, D.A.D., Squire, J.M.: J. Molec. Biol. 75, 33-55 (1973).
Pechere, J.-L., Capony, J.-P., Demaille, J.: Systematic Zoology 22, 533-548 (1973).
Perry, S.V., Cole, H.A., Head, J.F., Wilson, F.J.: Cold Spring Harbor Symp. Quant. Biol. 37, 251-262 (1972).
Potter, J.D.: Arch. Biochem. Biophys. 162, 436-441 (1974).
Potter, J.D.: Fed. Proc. 34, 2569 (1975a).
Potter, J.D.: Abstract, Biophys. Soc., p. 36a (1975b).
Potter, J.D., Gergely, J.: Biochemistry 13, 2697-2703 (1974).
Potter, J.D., Gergely, J.: J. Biol. Chem. 250, 4628-4633 (1975).
Potter, J.D., Seidel, J.C., Gergely, J.: Fed. Proc. 32, 1988 (1973).
Potter, J.D., Seidel, J.C., Leavis, P.C., Lehrer, S.S., Gergely, J.: In: Calcium Binding Proteins (eds. W. Drabikowski, H., Strzelecka-Golaszewska, E. Carafoli), pp. 129-152. Amsterdam: Elsevier 1974.
Regenstein, J.M., Szent-Györgyi, A.G.: Biochemistry 14, 917-925 (1975).
Solaro, R.J., Wise, R.M., Shiner, J.S., Briggs, F.N.: Circulation Res. 34, 525-530 (1974).
Spudich, J.A., Huxley, H.E., Finch, J.T.: J. Molec. Biol. 72, 619-632 (1972).
Tufty, R.M., Kretsinger, R.H.: Science 187, 167-169 (1975).
Van Eerd, J.P., Kawasaki, Y.: Biochem. Biophys. Res. Commun. 47, 859 to 865 (1972).
Van Eerd, J.P., Kawasaki, Y.: Biochemistry 12, 4972-4980 (1973).
Vibert, P.J., Haselgrove, J.C., Lowy, J., Paulson, F.R.: J. Molec. Biol. 71, 757-767 (1972).
Wakabayashi, T., Huxley, H.E., Amos, L.A., Klug, A.: J. Molec. Biol. 93, 477-497 (1975).
Weber, A., Herz, R.: J. Biol. Chem. 238, 599-605 (1963).
Weber, A., Murray, J.M.: Physiol. Rev. 53, 612-673 (1973).
Weeds, A., McLachlan, A.D.: Nature 252, 646-649 (1974).
Werber, M.M., Oplatka, A.: Biochem. Biophys. Res. Commun. 57, 823-830 (1974).
Wilkinson, J.M., Perry, S.V., Cole, H.A., Trayer, I.P.: Biochem. J. 127, 215-228 (1972).

Discussion

Dr. Weber: When you showed your data from which you came to the conclusion that two of the four Ca binding sites were essential for activating the actomyosin ATPase and that the sites where Ca and Mg could bind were non-essential, was this conclusion based on the fact that altering the Mg^{2+} concentration did not alter the Ca^{2+} range over which activation of the ATPase occurred?

Dr. Gergely: We calculated the dependence of the ATPase activity on (Ca^{2+}) for several sets of assumptions: (1) activity requires binding of Ca^{2+} to the high-affinity sites; (2) binding to all sites, (3) and binding to the low-affinity sites. Variation of (Mg^{2+}) would cause a change in activity for (1) and (2) but not for (3). The experimental data agree with (3). In the calculations we assumed that the parameters governing the binding of Ca^{2+} and Mg^{2+} to troponin also apply to the fibers. I should add that the changes in conformation I discussed occur in the Ca^{2+} concentration range corresponding to binding to the high-affinity sites. It seems to us that these high-affinity sites may be necessary for maintaining the proper configuration of troponin rather than being involved in regulation. The situation is perhaps similar to that obtaining with actin where tightly bound divalent metal ions seem necessary for maintaining the native structure.

Dr. Hess: Your CD data indicate a tremendous change in helicity. Can you correlate this with a helix part of the sequence of TnC? Can you correlate any part of the helix domain with this conformational change?

Dr. Gergely: The conformational change involves about half the total amount of helix content as predicted from the sequence data. The change in CD suggests that about half of the α-helix content is Ca^{2+}- or Mg^{2+}-dependent. This would correspond to 4 of the 8 α-helical regions whose existence has been deduced from the amino acid sequence data. One region, as I indicated, most likely to be involved in the Ca^{2+} dependent change is that including the single Cys residue of TnC and one of the cation binding sites.

Dr. Fisher: Do I understand correctly that you find no cooperative effect of Ca binding on your different binding sites?

Dr. Gergely: There is no cooperativity in Ca^{2+} binding but there is a cooperative effect if one looks at the ATPase activity. This is because you need more than one site occupied by Ca^{2+}, which produces apparent cooperativity in the effect on ATPase activity.

Dr. Weber: We have looked at Ca^{2+} binding to troponin and never found any positive cooperativity as long as myosin is absent.

Dr. Fisher: We have been looking at stripped muscle fibers and simultaneously measured binding of Ca to TnC phosphorylase kinase and tension. We always seem to see a slight positive cooperativity with a Hill factor of ca 1.6 both in the rabbit and the dogfish.

Dr. Gergely: How do you evaluate the cooperativity of binding?

Dr. Fisher: By the binding curves which are computer-fitted.

Dr. Weber: Is that in the presence of Mg and pure TnC?

Dr. Fisher: Yes. This was measured in the presence of 1 mM ATP, 2 mM free Mg, 30 mM P-creatine, 70 mM potassium, etc.

Dr. Weber: We have looked extensively at the binding of Ca in the complete system but have corrected only for troponin binding; but troponin bound to actin, tropomyosin and in isolated systems or in myofibrils and the binding is measured in the presence of sufficient ATP so that only negligible amounts of myosin are bound. The binding is invariably what you would describe as either two sites of binding or negative cooperativity without a trace of positive cooperativity.

Dr. Gergely: I should like to comment on the question of negative cooperativity. We have chosen to describe the Ca^{2+}-binding behaviour in terms of two high- and two low-affinity sites, but at present the experimental data would not permit a distinction between our model and one with negative cooperativity. Mathematically the curves would be indistinguishable. One might find a way to modify TnC and establish Ca^{2+} binding to one or more identifiable sites and thereby decide whether differences in binding strength are intrinsic or due to negative cooperativity.

Phosphorylation and the Regulation of the Function of Myosin and the Troponin Complex

S. V. Perry, H. A. Cole, N. Frearson, A. J. G. Moir, M. Morgan, and E. Pires

A. Introduction

There is now good evidence that three well-characterized enzymes pre-
cent in muscle can catalyse the phosphorylation by ATP of proteins of
the myofibril (Table 1). Myosin light chain kinase (Pires et al.,
1974) catalyses the transfer of one phosphate group to each of the

Table 1. Phosphorylation of the myofibrillar proteins

Protein	No. of sites phosphorylated per molecule
Contractile	
Actin	-
Myosin (Light Chain)	2
Regulatory	
Tropomyosin	-
Troponin C	-
Troponin I[a]	4 (2 major)
Troponin T[a]	3

[a] From rabbit white skeletal muscle.

18,000 to 19,000 dalton light chains of myosin from striated and smooth
muscle, the 'P light chain' (Frearson and Perry, 1975). Phosphorylase
kinase phosphorylates troponin I (Stull et al., 1972; Perry and Cole,
1973, 1974) and troponin T (Perry and Cole, 1973, 1974) and 3':5'-
cyclic AMP-dependent protein kinase acts mainly on troponin I (Stull
et al., 1972; Perry and Cole, 1973, 1974; cf. Pratje and Heilmeyer,
1972).

Protein phosphatases that will dephosphorylate myosin (Perry et al.,
1975a) and troponin (England et al., 1972; Cole and Perry, 1975) are
also present in muscle, thus enabling the phosphorylation-dephosphorylation processes to occur on the A and I filaments of the myofibril
under the control of Ca^{2+} and 3':5'-cyclic AMP.

B. Phosphorylation of Myosin

I. Myosin Light Chain Kinase

The number of light chain components of myosin that can be detected by
electrophoresis in sodium dodecyl sulphate depends upon the type of
muscle from which the myosin has been isolated. All the myosins that
have been studied so far possess a light chain of 18,000 to 20,000
molecular weight of which there are approximately two moles per molecule. This light chain is the substrate for myosin light chain kinase
which enzyme catalyses the transfer of the γ phosphate of ATP to a
serine residue in the light chain component (Fig. 1).

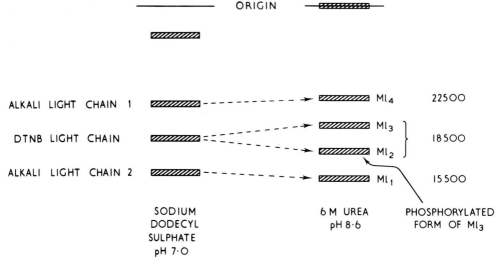

Fig. 1. Schematic representation of the electrophoresis of myosin from
white skeletal muscle of rabbit on polyacrylamide gel in 0.1% sodium
dodecyl sulphate, pH 7.0, and 6 M-urea, pH 8.6. Hatched areas indicate
bands staining with Brilliant Coomassie Blue. Figures on right: molecular weight

The enzyme was first identified in white skeletal muscle of the rabbit
in the course of following up the observation that the light chains of
myosin isolated from this tissue migrated on electrophoresis in 6 M-
urea at pH 8.6 as three or four bands, depending on the conditions of
extraction of the myosin (Perrie and Perry, 1970). The variation in
electrophoretic behaviour was due to the 18,500 molecular weight light
chain existing in two forms of slightly different electrophoretic mobility but apparently identical molecular weight, judging from their
lack of resolution on electrophoresis in sodium dodecyl sulphate. Analysis of the two forms of the 18,500 molecular weight component showed
that the form of higher electrophoretic mobility, at pH 8.6, component
Ml_2 (nomenclature of Perrie and Perry, 1970) was the phosphorylated
derivative containing 1 mol P/mol of the slower form component Ml_3.
A phosphorylated peptide was isolated from digests of component Ml_2
and its sequence shown to be Arg-Ala-Ala-Ala-Glu-Gly-Gly-[Ser-Ser(P)]-
Asn-Val-Phe (Perrie et al., 1973).

Although the original studies indicated that some phosphorylase kinase
preparations possessed myosin light chain kinase activity, purified
preparations of this enzyme and the 3':5'-cyclic AMP-dependent kinase,
both isolated from muscle, did not phosphorylate the 18,500 light chain
component. The enzyme that phosphorylates the myosin light chain has
now been isolated in a partially purified form from rabbit skeletal
muscle (Pires et al., 1974). The best preparations obtained to date
by a procedure involving isoelectric precipitation, ammonium sulphate
precipitation, chromatography on DEAE cellulose, DEAE Sephadex A50
and hydroxyapatite, transferred phosphate to the isolated total light
chain fraction of myosin at rates of 3-4 μmol P/mg enzyme/min. When
examined by electrophoresis at pH 8.6 the band with which the myosin
light chain activity was associated was judged by eye to represent not
more than about 10% of the total protein (Fig. 2). These results suggest that pure preparations of the myosin light chain kinase will be
extremely active and should be compared with phosphorylase kinase, the
best preparations of which have specific activities of about 10 μmole

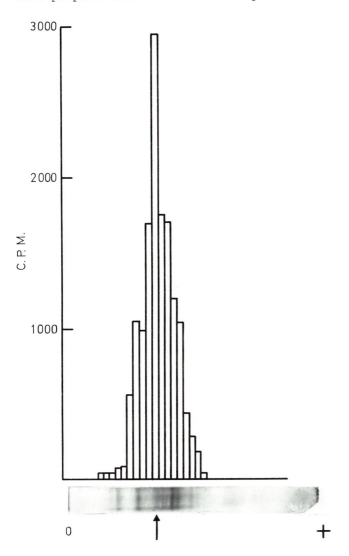

Fig. 2. Distribu-
tion of myosin
light chain kinase
activity after
electrophoresis.
Myosin light chain
kinase preparation
taken up to the
stage of chromato-
graphy on hydroxy-
apatite (100 μg)
of specific activ-
ity approx. 10 μ
moles P transfer-
red/mg protein/min
applied to 10%
polyacrylamide gel,
12.5 mM-Tris-40
mM-glycine, pH 8.6.
Gel slices in 2 mm
sections, protein
eluted by extrac-
tion with 1 ml
25 mM-Tris-HCl
buffer, pH 7.6,
2 mM-EDTA. O origin;
↑ band of myosin
light chain kinase.
Myosin light chain
kinase assayed on
20 μl of extract
as described by
Pires et al. (1974)

P/mg enzyme/min. Although the bulk of the myosin light chain kinase
is present in the sarcoplasmic protein fraction, myosin isolated by
conventional procedures is usually contaminated with the enzyme. This
can be demonstrated by incubating whole myosin with 12.5 mM-magnesium
acetate in 1 mM-dithiothreitol, O.1 mM-CaCl$_2$, 5 mM-ATP, 25 mM-Tris,
25 mM-glycerophosphate buffer, pH 7.6, at 25°C. Under these conditions
the 18,000-dalton light chain is phosphorylated and often is completely
converted into this form after ten minutes' incubation at 25°C. This
implies that unless special precautions are taken to purify myosin
most of the standard preparations used in the literature were probably
contaminated with the enzyme, and studies carried out on these preparations should be evaluated in this light.

In addition to Mg^{2+}, myosin light chain kinase requires Ca^{2+} for activity and the amounts of the latter cation normally contaminating
myosin preparations and the components of the incubation media are
adequate to activate the enzyme. As yet there is no evidence from the
relatively impure preparations tested that the enzyme has any requirement for 3':5'-cyclic AMP. The enzyme is highly specific for the light
chain components of 18,000-20,000 molecular weight present in myosin
from white skeletal muscle, this light chain is partially removed from
the myosin by extraction with 5'5'-dithiobis-2(-nitrobenzoic acid) and
has often been referred to as the 'DTNB light chain'. This property is
not, however, a universal property of this particular light chain. In
view of the fact that a light chain of 18,000 to 20,000 molecular
weight that can be phosphorylated is present in all myosins that we
have examined from muscles of higher animals and also in myosin from
platelets (Adelstein et al., 1973), we have suggested that it should
be designated the 'P light chain' (P for phophorylatable) (Frearson
and Perry, 1975). The enzyme does not phosphorylate troponin, phosphorylase b, casein or histone. As ITP, GTP and CTP do not act as donors,
the enzyme appears to be specific for ATP although ATPγS (kindly sup-
plied by Dr. F. Eckstein) will also act as phosphate donor, but less
effectively than ATP (Perry et al., 1975a).

The enzyme will phosphorylate the whole light chain fraction isolated
from myosin by the method of Perrie et al. (1973). With these preparations K$_m$ values at pH 7.6 and 25°C of 40-50 µM have been obtained.
The rate of phosphorylation of the 'P light chain' is not markedly
affected by removal from the myosin for similar rates are obtained
with a given concentration of light chains whether they have been previously isolated or intact in myosin (Fig. 3). Similar results were
obtained with actomyosin but there is preliminary evidence that the
rate of phosphorylation by crude kinase preparations may vary according
to the ratio of actin by myosin light chains present in the system.

As is the case with myosin from white skeletal muscle, cardiac myosin
isolated by extractants containing pyrophosphate and phosphate is also
isolated in the partially phosphorylated form. In cardiac myosin the
19,000-dalton component is the 'P light chain' and in most species
studied this light chain component is rapidly and spontaneously modified to a so-called 'satellite' form which has a slightly higher charge
(Frearson and Perry, 1975). A similar property has been reported for
the 'P light chain' of the myosin from rabbit white muscle (Perrie et
al., 1973). As is the case with both of these forms of the 'P light
chain', the so-called 19K and 19K' forms from cardiac myosin (Frearson
and Perry, 1975) can be phosphorylated either by the endogenous enzyme
of the cardiac muscle or myosin light chain kinase isolated from skeletal muscle (Fig. 4).

Using myosin light chain kinase partially purified from white muscle,
the whole light chain fractions of myosin from red and white skeletal

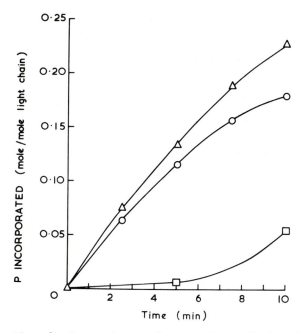

Fig. 3. Comparison of the rates of phosphorylation of myosin and the isolated light chain fraction by myosin light chain kinase. Myosin (5 mg/ml) or whole light chain fraction (0.75 mg/ml) isolated by method of Perrie et al. (1973) from rabbit white skeletal muscle was incubated with a myosin light chain kinase preparation (10 µg/ml) in 0.25 m-KCl, 50 mM-Tris-HCl buffer, pH 7.6, 1 mM-DTT, 0.02 mM-EDTA, 15 mM-Magnesium acetate, 0.1 mM-CaCl$_2$, 5 mM[γ-^{32}P]ATP (5 µCi/ml) at 25°C. At the time intervals indicated 0.15 ml was removed and solid urea added (0.5 g/ml). Each sample was applied to 8% polyacrylamide gel, 6 M-urea-20 mM-Tris-80 mM-glycine, pH 8.6. After electrophoresis and staining, slices of the polyacrylamide gel containing myosin light chains were dissolved in H$_2$O$_2$ (100 vol) at 80°C and radioactivity measured by the Cerenko method, as described by Perry and Cole (1973). △, myosin; o, light chains; □ , myosin without kinase

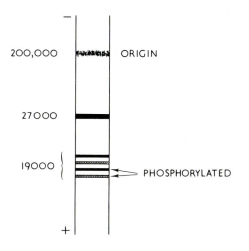

Fig. 4. Schematic representation of the band pattern obtained on electrophoresis of partially phosphorylated rabbit cardiac myosin in 8 M-urea, 20 mM-Tris-122 mM-glycine, pH 8.6. (From Frearson and Perry, 1975)

muscles were phosphorylated at very similar rates, whereas the light chain fraction of cardiac muscle was phosphorylated at about 30-40% of the rate obtained with the light chain fractions from skeletal muscle. The 20,000 molecular weight light chain component from myosin of smooth muscle of cow stomach and carotid artery and rabbit uterus was phosphorylated when incubated under the standard conditions with myosin light chain kinase preparations from white skeletal muscle (Frearson et al., 1975). This occurred with the whole myosin and the isolated light chains. Myosin light chain kinase is also present in red skeletal, cardiac and smooth muscle, and myosin isolated from these tissues, as in the case of myosin from white skeletal muscle, was usually contaminated with the endogenous enzyme. Although it cannot yet be decided whether the myosin light chain kinase enzyme is identical in the different muscle types, the 'P light chain' is clearly different but phosphorylatable in all cases. Thus phosphorylation of the 'P light chain' appears to be a feature of all striated and smooth muscles of the higher animals (Fig. 5).

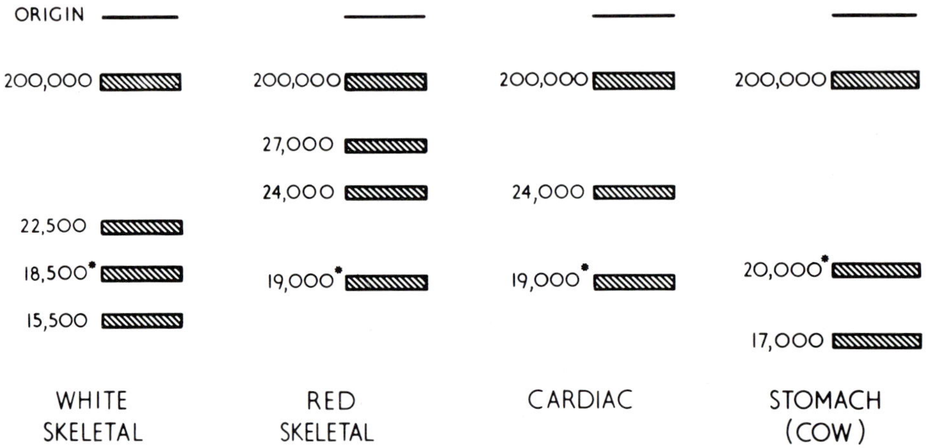

ORIGIN			
200,000	200,000	200,000	200,000
	27,000		
	24,000	24,000	
22,500			
18,500*	19,000*	19,000*	20,000*
15,500			17,000
WHITE SKELETAL	**RED SKELETAL**	**CARDIAC**	**STOMACH (COW)**

Fig. 5. Schematic representation of the light chains of myosin from different muscle types. The patterns obtained on electrophoresis in sodium dodecyl sulphate at pH 7.0 and their apparent molecular weights are illustrated. The band corresponding to a molecular weight of 18,000 to 20,000, marked an asterisk, is phosphorylated in each myosin type

II. Dephosphorylation of the 'P Light Chain' of Myosin

The original observations that if myosin was prepared in the absence of phosphate buffer it was rapidly dephosphorylated (Perrie and Perry, 1970) implied that muscle contained a phosphatase that rapidly dephosphorylated the 'P light chain'. In resting muscle myosin is fully phosphorylated and this persists mainly in this form for some hours post-mortem (Perrie et al., 1973). Once the muscle is homogenized in phosphate-free media, the 'P light chain' is rapidly dephosphorylated.

A protein phosphatase has been partially purified from the sarcoplasmic fraction of rabbit skeletal muscle that dephosphorylates phosphorylated myosin light chains at rates of 0.5-10 nmol/mg enzyme/min. Although the enzyme preparation probably contains the protein phosphatase as a minor component, its specific activities for phosphorylated myosin light chain fraction are comparable to those reported in the literature for preparations of other protein phosphatases. The most active preparations also dephosphorylated phosphorylase a and troponin previously phosphorylated by incubation with phosphorylase kinase, although at lower rates than obtained with phosphorylated myosin light chain as substrate.

As is the case with the myosin light chain kinase, the protein phosphatase is usually present in the standard three-times-precipitated myosin preparations. Extraction of rabbit myosin preparations with 4 mM-EDTA gave an extract with myosin light chain kinase and phosphatase activity (Perry et al., 1975a). On application of this extract to a column of DEAE cellulose equilibrated against 4 mM-EDTA, 25 mM-Tris-20 mM-HCl buffer, pH 7.6 and subsequent application of a gradient to 0.5 M-KCl, two peaks of myosin light chain kinase activity were eluted (Fig. 6). Elution of myosin light chain kinase activity in two peaks

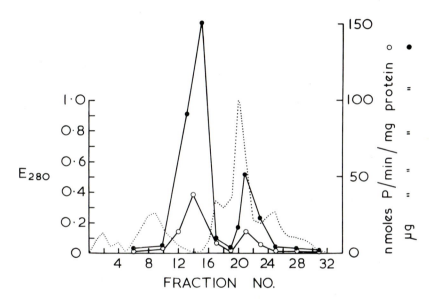

Fig. 6. DEAE cellulose chromatography of EDTA extract of freeze-dried rabbit skeletal muscle myosin. Extract of freeze-dried myosin in 25 mM-Tris-HCl buffer, pH 7.6, 4 mM-EDTA chromatographed on DEAE cellulose column (30 cm × 2 cm) equilibrated against buffer. Linear gradient applied to 0.5 M-KCl 25 mM-Tris-HCl buffer, pH 7.6, 4 mM-EDTA. --- E_{280}; o Myosin light chain kinase activity assayed in 0.1 mM-CaCl$_2$, 12.5 mM-magnesium acetate, 25 mM-Tris-HCl buffer, pH 7.6, 5 mM-[γ-^{32}P]ATP, 0.2 ml eluate. Incubated 5 min at 25°C. Reaction stopped with 1/3 vol of 15% trichloroacetic acid. Precipitate washed 3 times with 5% trichloroacetic acid and ^{32}P estimated as described by Perry and Cole (1973); ● "ATPase" activity measured in 25 mM-Tris-HCl buffer, pH 7.6, 30 µg whole light chains from rabbit fast skeletal muscle, 5 mM-Tris ATP, 0.2 ml column eluate. Incubated 5 min at 25°C and reaction stopped with 1/3 vol. 15% trichloroacetic acid. Inorganic phosphate measured by the method of Fiske and Subbarow (1925). (From Perry et al., 1975a)

was a consistent observation when crude myosin light chain preparations were chromatographed under these conditions, and may reflect the existence of the enzyme in two forms. Purer samples of the enzyme did not behave in this way. Myosin light chain phosphatase activity was also eluted in the two fractions that possessed myosin light chain kinase

activity. This suggests some similarity of properties or possibly interaction between the two enzymes. Myosin light chain phosphatase activity could not be demonstrated in more highly purified preparations
of myosin light chain kinase.

The combined action of these two enzymes in the presence of a trace of
the 'P light chain' results in a highly specific, Ca^{2+}-sensitive hydrolysis of ATP.

$$\text{ATP} + \text{'P light chain'} \xrightarrow[\text{Mg}^{2+} + \text{Ca}^{2+}]{\text{kinase}} \text{Phosphorylated 'P light chain'} + \text{ADP}$$

$$\text{Phosphorylated 'P light chain'} \xrightarrow[\text{Mg}^{2+}]{\text{Phosphatase}} \text{'P light chain'} + \text{P}_i$$

So long as the two enzymes are present, which indeed they are in muscle
and usually as contaminants in normal myosin preparations, only a trace
of 'P light chain' is required, i.e. in the free form or associated
with myosin. Thus when 30 µg of the whole light chain fraction of myosin
is added to fractions with myosin light chain kinase activity obtained
by chromatography of myosin extracts on DEAE cellulose, ATPase activity
can be demonstrated (Fig. 6). It is unlikely that sufficient amounts
of the enzymes concerned are present to give very high 'ATPase' activity of this type in normal myosin preparations. Nevertheless, in systems
in which Ca^{2+} regulation of myosin ATPase is described in the absence
of the troponin system, it is necessary to confirm that Ca^{2+} sensitivity is not due to the coupled system described. If such activity is
present, removal of the 'P light chain' would stop the Ca^{2+}-sensitive
'ATPase' due to the combined action of the two enzymes.

III. Function of Phosphorylation of Myosin

The highly specific phosphorylation of a single site present in the
two 'P light chains' of the myosin molecule, i.e. one site for each
globular head, implies a unique function of this process. The presence
also of a protein phosphatase to remove the group indicates that a
phosphorylation-dephosphorylation cycle could occur. The phosphorylation will be subject to regulation because of the requirement of myosin
light chain kinase for Ca^{2+} at concentrations similar to those which
activate contraction. Thus phosphorylation can only occur when the muscle is activated, but so far nothing can be said about the factors
controlling dephosphorylation. The phosphorylation site must clearly
be readily available on the intact myosin molecule, for the 'P light
chain' is as readily phosphorylated whether isolated or associated with
myosin. In this respect, the recent report (Collins, 1975) that the
site is close to the N-terminal of the light chain is of some significance.

Despite an extensive investigation of the enzymic properties, no marked
differences have been observed between the properties of phosphorylated
and dephosphorylated myosin. The properties studied were the Mg^{2+}-,
Ca^{2+}- and K^+-activated ATPase activities of myosin, heavy meromyosin
(specially prepared to preserve the light chains intact) acto-heavy
meromyosin and desensitized actomyosin. Also no clear differences in
actin-binding characteristics could be observed from ATPase studies
using double reciprocal plots (Eisenberg and Moos, 1967), viscosity
or precipitation studies. Phosphorylation did not markedly change the
Ca^{2+} binding properties of the 'P light chain'.

Although these studies do not reveal evidence of changes in the actin-binding properties of myosin, preliminary evidence, to which brief mention has been made, does suggest that actin can interact with the 'P light chain' to activate or inhibit its phosphorylation, depending on the molar ratio of actin to 'P light chain'. The significance of these effects cannot yet be evaluated but they do suggest that the 'P light chain' may have a special function in the interaction of myosin with actin.

Although the turnover number of the myosin light chain kinase for ATP is probably between one and two orders of magnitude greater than that of myosin, the concentration in muscle is much less and the total capacity for ATP turnover per gram of muscle is 50 times greater in the case of myosin. This would imply that if the phosphorylation of the light chain has a regulatory role, it is probably not synchronous with the cross-bridge cycle but represents a process superimposed upon it and exerting its effect over a number of cycles.

C. Phosphorylation of the Components of the Troponin Complex

The evidence from a number of laboratories is that troponin I can be phosphorylated by 3':5'-cyclic AMP-dependent protein kinase and by phosphorylase kinase. Troponin T is phosphorylated by phosphorylase kinase and possibly also by 3':5'-cyclic AMP-dependent protein kinase (Fig. 7), although in our laboratory the rate of phosphorylation of

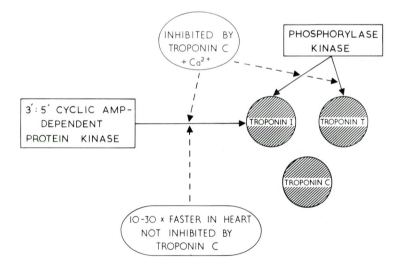

Fig. 7. Phosphorylation of the components of the troponin complex

troponin T by 3':5'-cyclic AMP-dependent protein kinase is very slow (cf. Pratje and Heilmeyer, 1972). Although it is difficult to compare rates of phosphorylation obtained in different laboratories, the apparent differences may be due to the difference in specificity between the 3':5'-cyclic AMP-dependent protein kinases of skeletal and cardiac muscle. With the exception of the rates of phosphorylation obtained with 3':5'-cyclic AMP-dependent kinase using cardiac troponin I as substrate, the rates of phosphorylation of troponin I and troponin T by both enzymes are rather low. Certainly the rates of phosphorylation obtained in our laboratory with phosphorylase kinase are 1% or less than those obtained with phosphorylase b as substrate. In the case of troponin I considerable phosphorylation by phosphorylase kinase occurs in the presence of EGTA, which chelator completely inhibits the phosphorylation of phosphorylase b by the enzyme. The phosphorylase kinase preparations also vary in their relative phosphorylating activity towards troponin and phosphorylase b. Further, in I strain mice, which do not possess active phosphorylase kinase, the troponin T is phosphorylated when isolated from skeletal muscles of these mice (Cole, Cohen and Perry, unpublished observations). These results throw some doubt on whether phosphorylase kinase is an effective phosphorylating enzyme for the troponin complex *in vivo* and imply that another uncharacterized kinase in muscle may phosphorylate certain sites on the complex. On the other hand, the phosphorylation of troponin I by 3':5'-cyclic AMP-dependent protein kinase would appear to be specific and of particular significance in cardiac muscle.

Although the *in vitro* studies suggest that in some cases at least, the rates of phosphorylation of troponin by well-characterized enzyme systems for muscle are low, troponin I and troponin T are phosphorylated *in vivo*. When isolated as the complex from rabbit skeletal muscle troponin contains about 1 mole of phosphate per 80,000 g. This phosphate is virtually all associated with an N-terminal peptide of troponin T which appears to be particularly stable to the endogenous protein phosphatases. The troponin complex isolated from cardiac muscle contains significantly more covalently bound phosphate amounting to 2-3 mol per 80,000 g. On subsequent fractionation of cardiac troponin by the standard procedures, covalently bound phosphate is present in both troponin I and troponin T (Cole and Perry, 1975). If troponin I is isolated from fresh muscle by an affinity chromatographic procedure that permits its rapid isolation in a pure form from a homogenate of the fresh muscle in 8 M-urea by a single-step procedure (Syska et al., 1974), it is found to be phosphorylated. In the case of white skeletal muscle troponin the phosphate content is about 0.5 mole/mole but values up to 2-3 mole/mole have been obtained with rabbit cardiac troponin I. Thus in the case of troponin I at least, and particularly in cardiac muscle (Cole and Perry, 1975), effective phosphorylation-dephosphorylation systems would appear to exist for the troponin system.

As yet a role for phosphorylation of the components of the troponin complex has not been demonstrated. Preliminary experiments indicate that there is no significant difference in the inhibition of the Mg^{2+}-stimulated ATPase of desensitized actomyosin by fully phosphorylated and dephosphorylated troponin from skeletal muscle (Perry and Cole, 1974). It is, however, questionable whether this type of experiment carried out under steady-state conditions, would reveal any striking effects if the role of phosphorylation is to modify the interaction between the troponin components that occurs during the dynamic phase of the contractile cycle.

I. Phosphorylation and Interactions between the Components of the Troponin Complex

At the moment there is no clear understanding of the functional significance of phosphorylation of the troponin components. Nevertheless the fact that the phosphorylation occurring at defined sites on the molecules of troponin I and troponin T is modified in the presence of troponin C (Perry and Cole, 1974) can be used to acquire information about the sites of interaction between the components of the complex.

The major site of phosphorylation of troponin I from rabbit white
skeletal muscle by phosphorylase kinase is a threonine residue at position 11 and the corresponding site for 3':5'-cyclic AMP-dependent
protein kinase is serine 118 (Moir et al., 1974; Huang et al., 1974).
Troponin C forms a complex with troponin I that is much strengthened
by Ca^{2+} (Head and Perry, 1973). The formation of the complex almost
completely inhibits phosphorylation by phosphorylase kinase and 3':5'-
cyclic AMP-dependent kinase, indicating that the phosphorylation sites
at threonine 11 and serine 118 are no longer available to these enzymes
(Fig. 8. In both cases the inhibition of phosphorylation produced by
troponin C is complete when equimolar amounts of troponin C and troponin

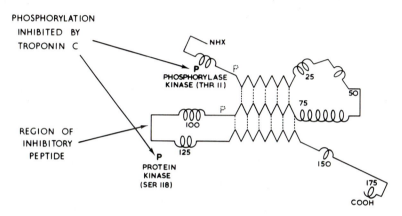

Fig. 8. Phosphorylation sites of troponin I from rabbit white skeletal
muscle. Major and minor phosphorylation sites are indicated by a bold
and dotted letter P respectively. The conformation is that predicted
by the application of the Chou and Fasman (1974) procedure to the primary sequence determined by Wilkinson and Grand (1975)

I are present (Fig. 9). These facts are relevant to the mode of action
of troponin C in neutralising the inhibitory activity of the Mg^{2+}-
stimulated ATPase of desensitized actomyosin.

It has been shown (Syska et al., 1975; Perry et al., 1975b) that the
region of the troponin I molecule consisting of residues 96 to 117
interacts with actin and is responsible for the inhibitory activity of
the whole molecule. This region is adjacent to the serine residue at
position 118, that is the main site of phosphorylation by 3':5'-cyclic
AMP-dependent protein kinase. Thus from the studies of the phosphorylation of troponin I and of the interaction of peptide fragments of
troponin I with actin it can be concluded that the interaction with
troponin C has effects on two regions of the troponin I molecule, one
of which is very close to that involved in interaction with actin. Two
possible mechanisms to explain the observations present themselves.
In the first it is postulated that troponin C interacts with troponin
I in the N-terminal region, hence blocking directly the phosphorylation
of threonine 11. By inducing a conformational change in the troponin I
molecule the interaction with troponin C at the N-terminal region leads
to changes in the region of the polypeptide chain represented by residues 96-118. In consequence, the serine residue at position 118 becomes unavailable for phosphorylation and interaction with actin cannot
occur; inhibition of the Mg^{2+}-stimulated ATPase of actomyosin can no
longer take place. In this scheme the two sites of phosphorylation are

118

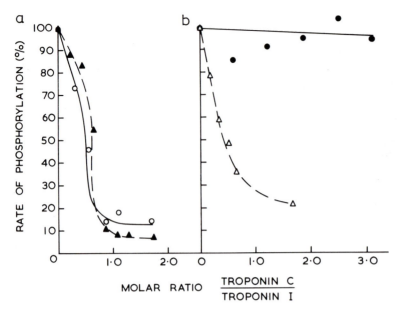

Fig. 9a and b. Effect of troponin C on the phosphorylation of troponin I from skeletal and cardiac muscle. Conditions for phosphorylation as described by Cole and Perry (1975). (a) Rabbit troponin (0.1-0.15 mg/ml) incubated with phosphorylase kinase (0.025 mg/ml), [γ-³²P]ATP (1 mM; 1 μCi/ml) and troponin C. ▲ Troponin I and troponin C from white skeletal muscle. O Troponin I and troponin C from cardiac muscle. (b) Rabbit troponin I (0.1-0.15 mg/ml) incubated with bovine cardiac 3':5'-cyclic AMP-dependent protein kinase (0.01-0.025 mg/ml), [γ-³²P]ATP (1 mM; 1.6 μCi/ml) and troponin C. △ Troponin I and troponin C from white skeletal muscle. ● Troponin I and troponin C from cardiac muscle

widely separated on the surface of the molecule, whereas in the alternative hypothesis, the sites are adjacent on the surface of the troponin I. In the latter situation interaction with troponin C blocks both phosphorylation sites and at the same time renders the actin-binding site unavailable. In this way it neutralises the inhibitory activity of troponin I.

In both schemes it is postulated that the N-terminal region of troponin I has special significance for the binding of troponin C. Studies with troponin I from white skeletal muscle of the rabbit do not permit this conclusion but investigations on the effects of troponin C on the phosphorylation of cardiac troponin I (Cole and Perry, 1975) suggest this may be the case. In the case of cadiac troponin I, troponin C inhibits phosphorylation by phosphorylase kinase but not that catalysed by the 3':5'-cyclic AMP-dependent protein kinase (Fig. 9). If it is presumed that the sites of phosphorylation on cardiac troponin I by the two enzymes are similar to those on white muscle troponin I, the observation suggests that interaction with troponin C at the N-terminal site may be a common feature of all forms of troponin I.

The observation also throws some light on the structure-function differences between troponin I of skeletal and cardiac muscles, but final conclusions about its significance must await more knowledge of the primary sequence of cardiac troponin I. One can conclude that in white muscle troponin I at least, each of the two main phosphorylation sites

are located at or very close to the regions of the polypeptide chain
that are involved in interaction with troponin C and actin.

The nature of the sites of phosphorylation on the troponin T molecule
is less well defined. There are probably three in troponin T from rabbit white skeletal muscle that are phosphorylated by phosphorylase
kinase, albeit somewhat slowly. They are: a serine site close to the
N-terminal which is relatively stable in the phosphorylated form and
which is usually phosphorylated in troponin T after isolation; a serine
site adjacent to the second methionine residue from the N-terminal,
approximately at position 160; and a third not yet characterized site
in the C-terminal half of the molecule. Troponin T also interacts with
troponin C to form a complex. As is the case with troponin I, phosphorylation of troponin T is inhibited by troponin C and the inhibition
is complete when equimolar amounts of the two proteins are present.
It has not yet been possible to decide if phosphorylation at any particular site is preferentially inhibited. It is worthy of note that
one of the phosphorylation sites on troponin T is adjacent to the region of the molecule shown to be involved in interaction with tropomyosin, namely the region of the molecule running from residue 70-160
(Jackson et al., 1975).

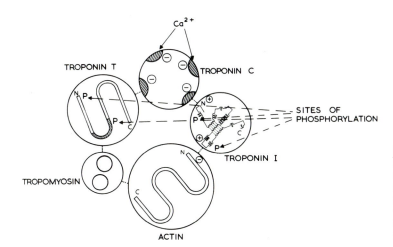

Fig. 10. Schemat-
ic representation
of the interac-
tions between the
components of the
troponin complex,
actin and tropo-
myosin

Certain general conclusions can be drawn about the interactions of the
components of the troponin complex and the relation of phosphorylation
to them. These are presented in the scheme illustrated in Fig. 10. In
the case of troponin I, the regions of polypeptide chain which the
evidence suggests are involved in interaction with troponin C and actin
are strongly basic and carry net positive charges of 7 and 8 respectively. Major sites of phosphorylation are localized close to or even
form part of the interaction sites. Thus phosphorylation-dephosphorylation reactions could produce significant changes in net charge at these
sites and change the stength of interactions and the behaviour of the
system *in vivo*. In this respect it is perhaps significant that there
is no evidence that troponin C, which is a strongly acidic protein,
can be phosphorylated. Addition of negative charges to a molecule already strongly negatively charged would not be expected to have such
a marked effect on its interaction with the basic proteins, troponin I
or troponin T. In the case of cardiac troponin I the evidence for a
specific role for phosphorylation is even more striking, for two or
three relatively labile phosphate groups can be transferred rapidly to

the protein by the action of 3':5'-cyclic AMP-dependent protein kinase. This process could play a part in the inotropic response of cardiac muscle to the catecholamines. The recent findings of England (1975) that the increase in the force of contraction of cardiac muscle is parallelled by the extent of phosphorylation of the troponin I in response to adrenaline gives support to this view.

D. Conclusions

Despite the clear evidence for specific sites of phosphorylation both on the myosin component of A filament and on the troponin I and troponin T components of the I filament it is not yet possible to present a role that is supported by good direct experimental evidence for a phosphorylation-dephosphorylation process at any of the sites discussed. The evidence is presumptive but the pattern which emerges in all cases is that phosphorylation probably modulates the interaction of the components involved by changing the net charge at those regions of the molecules in close contact. Protein interactions are basic to the molecular mechanism of contraction. During the contraction interactions involving myosin and actin, troponin C and troponin I, troponin I and actin, troponin T and tropomyosin occur probably sequentially and in a cyclical manner within milliseconds. In every case mentioned there is a site close to the region involved in the interaction that can be phosphorylated by enzymes present in muscle. The phosphorylations are regulated by either Ca^{2+} or 3':5'-cyclic AMP, two of the major agents for the regulation of intracellular function. It would seem unreasonable to ascribe these correlations and associations to pure chance. It is more likely that our current inability to identify with certainty the processes involved merely reflects the relative lack of discrimination of the techniques currently employed to study the integrated activity of the components of the contractile system.

References

Adelstein, R.S., Conti, M.A., Anderson, W.: Proc. Nat. Acad. Sci. U.S.A. 70, 3115-3119 (1973).
Chou, P.Y., Fasman, G.D.: Biochemistry 13, 222-245 (1974).
Cole, H.A., Perry, S.V.: Biochem J. (in press) (1975).
Collins, J.H.: Fed. Proc. 34, 539 (1975).
Eisenberg, E., Moos, C.: J. Biol. Chem. 242, 2945-2951 (1967).
England, P.: FEBS Lett. 50, 57-60 (1975).
England, P.J., Stull, J.T., Krebs. E.G.: J. Biol. Chem. 247, 5275-5277 (1972).
Frearson, N., Focant, B., Perry, S.V.: FEBS Lett. (in press) (1975).
Frearson, N., Perry, S.V.: Biochem. J. (in press) (1975).
Head, J.F., Perry, S.V.: Biochem. J. 137, 145-154 (1974).
Huang, T.S., Bylund, D.B., Stull, J.T., Krebs, E.G.: FEBS Lett. 42, 249-252 (1974).
Jackson, P., Amphlett, G.W., Perry, S.V.: Biochem. J. (in press) (1975).
Moir, A.J.G., Wilkinson, J.M., Perry, S.V.: FEBS Lett. 42, 253-256 (1974).
Perrie, W.T., Perry, S.V.: Biochem. J. 119, 31-38 (1970).
Perrie, W.T., Smillie, L.B., Perry, S.V.: Biochem. J. 135, 151-164 (1973).
Perry, S.V., Amphlett, G.A., Grand, R.J.A., Jackson, P., Syska, H., Wilkinson, J.M.: Symposium on Calcium Transport in Contraction and Secretion. Bressanone, May 1975 (1975b).

Perry, S.V., Cole, H.A.: Biochem. J. $\underline{131}$, 425-428 (1973).
Perry, S.V., Cole, H.A.: Biochem. J. $\underline{141}$, 733-743 (1974).
Perry, S.V., Cole, H.A., Morgan, M., Moir, A.J.G., Pires, E.: Proc.
  9th FEBS Meeting Proteins of Contractile systems (ed. E.N.A. Biro)
  Vol. 31. Budapest: Akademai Kiado; Amsterdam: North Holland (1975a).
Pires, E., Perry, S.V., Thomas, M.A.W.: FEBS Lett. $\underline{41}$, 292-296 (1974).
Pratje, E., Heilmeyer, Jr., L.M.G.: FEBS Lett. $\underline{27}$, 89-93 (1972).
Stull, J.T., Brostrom, C.O., Krebs, E.G.: J. Biol. Chem. $\underline{247}$, 5272-5274
  (1972).
Syska, H., Perry, S.V., Trayer, I.P.: FEBS Lett. $\underline{40}$, 253-257 (1974).
Syska, H., Wilkinson, J.M., Grand, R.J.A., Perry, S.V.: Biochem. J.
  (in press) (1975).
Wilkinson, J.M., Grand, R.J.A.: Biochem. J. (in press) (1975).

Discussion

Dr. Helmreich: Is the myosin light chain kinase itself phosphorylat-able?

Dr. Perry: I wish I could be certain about that. We believe the enzyme may be phosphorylated but we need to look at this aspect more carefully.

Dr. Helmreich: What about the specificity of the corresponding phos-phatase?

Dr. Perry: It is not as specific as I would like. Some preparations will dephosphorylate phosphorylated myosin light chains much faster than they dephosphorylate phosphorylase $\underline{a}$. I would like to believe that the protein phosphatase associated with myosin is not identical with phosphorylase phosphatase, but as you know the protein phosphatases in general are not well characterised.

Dr. Hamprecht: You mentioned that your myosin light chain kinase was not activated by cyclic AMP. I am wondering whether other fractions might contain an inhibitory subunit of this protein kinase. If this fraction were added, the protein kinase might be inhibited and then reactivated by cyclic AMP added subsequently?

Dr. Perry: That is a very pertinent comment. Most of our work so far has been carried out on relatively impure preparations. The purest preparations obtained by preparative electrophoresis often contain two bands when examined by electrophoresis on sodium dodecyl sulphate. We are not sure whether both components are part of the enzyme.

Dr. Fischer: Did you mention the molecular weight of the single band of myosin light chain kinase?

Dr. Perry: On electrophoresis in sodium dodecyl sulphate a consistent component of our best preparations is a band corresponding to a mole-cular weight of 70,000-80,000.

Dr. Fischer: Is the presence of Ca mandatory for activity?

Dr. Perry: It appears to be essential in all the preparations we have examined so far.

Dr. Weber: Does the phosphorylation of TNT influence the binding of this component to tropomyosin?

Dr. Perry: We have not yet studied this aspect.

Myosin-linked Calcium Regulation

J. Kendrick-Jones

A. Introduction

It is generally accepted that all muscles contract by a relative sliding
of the actin and myosin filaments past each other and that regulation
of this contractile activity is controlled by the concentration of
calcium ions within the muscle. The ability of the actin and myosin
filaments to respond to changes in the calcium ion concentration is
due to the presence of specific proteins associated with these filaments. In vertebrate striated muscles, the calcium regulatory protein,
troponin, which also binds calcium, is associated with the actin filaments (Ebashi and Endo, 1968) (Fig. 1). In these muscles, tropomyosin
is required for regulation, its role being to transmit the calcium-
induced changes which occur in the troponin complex to affect the active site on the actin. In the absence of calcium, troponin, via tropomyosin "switches off" the active site of actin, thus preventing inter-

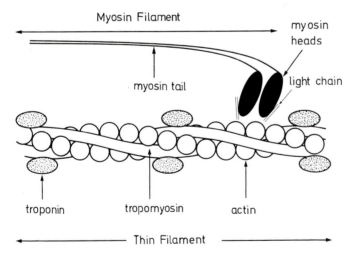

Fig. 1. The control of muscular contraction. Diagrammatic representation of myosin cross-bridge and thin filament, showing the 'positions'
of the calcium regulatory components. The cross-bridge consist of two
globular 'heads' containing the enzymic and actin-combining sites. Associated with these 'heads' are small polypeptides, 'light chains',
which in molluscan muscles act as calcium regulatory subunits. The
troponin complex is localised at every 38.5 nm along the thin filament
in vertebrate skeletal muscle. It is attached to the tropomyosin which
lies in each groove of the actin filament. The troponin complex consists
of equimolar amounts of the calcium binding component (TN-C), the inhibitory protein (TN-I) and the 'tropomyosin binding' component (TN-T)

action with the myosin cross-bridge, and on addition of calcium, "switches" it on again (Ebashi and Endo, 1968).

In molluscan and a number of "primitive" invertebrate muscles, troponin appears to be absent and although tropomyosin is present on the actin filaments, it is not required for regulation (Lehman et al., 1972). In these muscles, calcium regulation is associated with myosin; molluscan myosin binds calcium and its interaction with pure actin (unregulated) is dependent on the calcium ion concentration. In the absence of calcium, the active sites on the myosin are blocked, thus preventing interaction with actin (Kendrick-Jones et al., 1970).

In molluscan muscles, regulation by calcium is dependent on the presence of specific low molecular weight polypeptides on the myosin - 'light chains' (Fig. 1). These light chains are called the EDTA light chains, since they are selectively removed from myosin if the divalent cation concentration is lowered by the addition of EDTA, and their release results in a loss of regulation. The myosin is then desensitized and calcium is no longer required for interaction with actin. In addition the amount of calcium bound is reduced (Szent-Györgyi et al., 1973).

Recombination of the EDTA light chain with the desensitized myosin in the presence of divalent cations restores calcium regulation and calcium binding, showing that the chain functions as a regulatory subunit.

I. Myosin-linked Regulation in Molluscan Muscles

1. Subunit Structure of Molluscan Myosins

Scallop myosin contains two distinct types of light chains: the EDTA light chain which is required for calcium regulation, and the "thiol-containing" light chain. The thiol light chain is only released when the myosin is denatured, and is therefore believed to be analogous to the essential "alkali" light chains of rabbit myosin, which have been implicated in the control of the enzymatic activity of the myosin (Dreizen and Gershman, 1970). Both types of scallop light chains have similar molecular weights (they give a single rather diffuse band at 17,000 on SDS acrylamide gels, Fig. 2) but they may be separated on the basis of charge (two distinct light chain bands are obtained on gel electrophoresis in the presence of urea, Fig. 2). Each myosin molecule contains two moles of the EDTA light chain and two moles of the thiol light chain (Kendrick-Jones et al., 1975). The proteolytic fragments of myosin, heavy meromyosin, which contains both myosin heads joined by a short tail, and subfragment-1 (single myosin head) indicate that each myosin head contains one of each type of light chain, i.e. one EDTA and one thiol light chain per active site. A survey of different molluscan muscles (Kendrick-Jones et al., 1975) has shown that all the myosins contain two moles of calcium regulatory light chain and two moles of "alkali" light chain, analogous to the EDTA and thiol light chains of scallop myosin.

II. Role of the EDTA Light Chain in Calcium Regulation

1. Myosin-actin Interaction

The interaction of molluscan myosins with unregulated rabbit actin (actin free of the troponin components) is dependent on the calcium ion concentration. That is, in the absence of calcium, the Mg^{2+}-dependent ATPase of mollusc myosin is not activated by actin (Table 1); the myosin is 'switched off'. If the myosin is treated with high con-

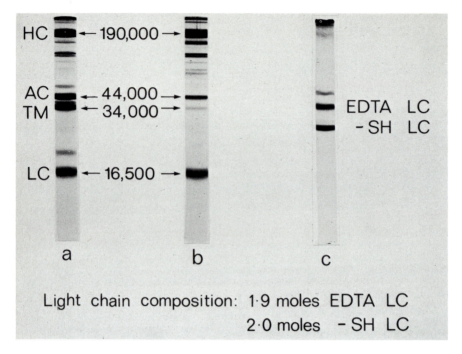

Fig. 2a-c. Subunit composition of scallop myosin and myofibrils revealed by acrylamide gel electrophoresis. (a) SDS-acrylamide gel (10%)
of scallop myofibrils. (b) SDS-acrylamide gel (10%) of scallop myosin.
(c) 8 M urea-acrylamide gel (10%) of scallop myosin. Gels stained with
Coomassie brilliant blue. Light chain composition established by densitometry of 8 M urea-acrylamide gels of scallop myosin (Kendrick-
Jones et al., 1975) HC, heavy chains of myosin; AC, actin; TM, tropomyosin; LC, light chains of myosin

centrations of EDTA (10 mM) one of the two EDTA light chains is released and as a result the myosin loses its ability to be regulated
by calcium. It is desensitized and its actin-activated ATPase in the
absence of calcium is the same as that in the presence of calcium. It
is remarkable that the removal of only half the EDTA light chains should
lead to a complete loss of calcium regulation. The isolated EDTA light
chain readily recombines with the desensitized myosin in the presence
of divalent cations and calcium regulation is restored (Table 1, Fig.
5).

The light chain may also be removed from myofibrils and actomyosin in
the presence of similar concentrations of EDTA. Only 50% of the light
chains are released although calcium regulation is again completely
abolished. The light chain recombines with desensitized myofibrils,
regardless of whether they are relaxed or "in rigor", with a resultant
full recovery of calcium regulations. It therefore appears that the
light chain binding sites on the myosin are accessible regardless of
the conformational state of the myosin (nucleotide free, steady-state
intermediate, myosin ADP* P, or other intermediate state) or whether
the myosin is combined with actin.

Table 1. The effect of the release and recombination of the EDTA light chain on the Mg^{2+}-dependent ATPase and calcium binding of scallop myosin

| | EDTA light chain content (moles myosin^{-1}) | Actin-activated Mg^{2+} ATPase (mol ATP s^{-1} per site) | | Calcium binding Number of sites | Binding constants |
		10^{-8}M Ca^{2+}	10^{-4}M Ca^{2+}		(M^{-1})
Myosin	1.86	0.015	0.94	1.8	4×10^6
'Desensitized' myosin	0.96	0.680	0.72	1.0	2.7×10^6
'Desensitized' myosin + EDTA light chain	1.72	0.011	0.79	1.7	3.2×10^6
EDTA light chain	-	-	-	~0.4	$\sim 10^5$

Myosin was desensitized by treating with 10 mM EDTA (Szent-Györgyi et al., 1973). EDTA light chain content determined by densitometry of 8 M urea-acrylamide gels stained with Coomassie brilliant blue. Actin-activated magnesium-dependent ATPase activity, using 0.05 -0.1 mg ml^{-1} myosin and 0.02-0.03 mg ml^{-1} rabbit actin, measured in 0.5 mM ATP, 25 mM NaCl, 2 mM $MgCl_2$ and either 0.1 mM EGTA or 0.1 mM $CaCl_2$ in a radiometer pH stat at pH 7.6 and 25°. Calcium binding measurements on the myosin were determined by the procedure outlined by Szent-Györgyi et al. (1973) and on the isolated EDTA light chain using equilibrium dialysis. There is considerable heterogeneity in the calcium binding data, therefore the values for the association constants are approximate.

2. Calcium Binding

In the presence of magnesium (1 mM), scallop myosin binds about two moles of calcium (we assume that each myosin head contains one binding site) with an association constant (about 10^6 M^{-1}) which is closely comparable to the association constant of the calcium binding component (troponin-C) of the troponin complex (Greaser et al., 1972) (Table 1). When the myosin is 'desensitized' by treatment with EDTA, one of the EDTA light chains is released and the amount of calcium bound is decreased by about 40%. This decrease in bound calcium is due to the loss of a binding site and not to an altered affinity, since the binding constant of the desensitized myosin is close to that of intact myosin (Szent-Györgyi et al., 1973).

Where are the calcium binding sites on the myosin? They appear to be located on the myosin 'heads', since subfragment - 1, the single myosin head, produced by proteolytic cleavage of myosin, contains about 0.6 calcium binding sites. However the exact location of the sites on the myosin 'head' is in doubt. We do not know whether they are on the light chain or heavy chain or require a combination of heavy chain and light chain. The correlation between the removal of one EDTA light chain and the loss of one calcium binding site suggests that the binding site is on the light chain. However, calcium binding measurements on isolated EDTA light chains suggest that the amount of calcium bound is rather low (Szent-Györgyi et al., 1973). Recently, using a rapid equilibrium dialysis procedure, we observed measurable calcium binding on the isolated light chains in the presence of 0.1 mM magnesium (Jakes and Kendrick-Jones, unpublished observations) (Table 1). One reason for the

difficulty in detecting calcium binding on isolated EDTA light chains
may be their anomalous behaviour; these light chains show time-dependent
aggregation in the analytical ultracentrifuge, they are excluded from
G75 Sephadex (only proteins with molecular weights higher than about
70,000 should be excluded) and they have a strong tendency to aggregate
on ion exchange columns. However they are obviously not irreversibly
'denatured' since they will still combine with desensitized scallop
myosin in stoichiometric amounts and restore calcium regulation (Table
1, Fig. 5). An explanation for the behaviour of the isolated EDTA light
chains is now being sought, which may resolve the question of the location of the calcium binding sites on molluscan myosin.

The EDTA light chains are tightly bound to scallop myosin in the presence of divalent cations, about 10^{-4} M magnesium or 10^{-6} M calcium
(Szent-Györgyi et al., 1973). The fact that high concentrations of
EDTA (5-10 mM) are required to remove 50% of this light chain suggests
that either divalent cations are involved directly in binding the light
chain to the myosin heavy chain or that possibly the light chain is
released as a consequence of some 'structural' change in the myosin
head induced by the removal of tightly-bound divalent cation, probably
magnesium. It is interesting that in other molluscan myosins, such as
Spisula (surf clam) and *Loligo* (squid) species, EDTA removed only a
very small proportion of the regulatory light chain (Kendrick-Jones et
al., 1975). In these myosins, no effective method has been found for
removing the regulatory light chains without destroying the enzyme activity of the myosin.

3. Why is Only One EDTA Light Chain Released?

The release of only half the total EDTA light chain complement might
be explained if there are two types of this calcium regulatory subunit.
If, however, the light chain released with EDTA is compared to the remaining regulatory light chain, which can be selectively released from
the desensitized myosin by treating with the Ellman reagent 5,5'-
dithiobis-(2-nitrobenzioc acid) DTNB (a treatment which irreversibly
denatures scallop myosin) their amino acid analyses and tryptic peptide
maps are identical (Fig. 3). Since the two 'EDTA' light chains in a
myosin molecule appear to be chemically identical, one might ask if the
release is a 'random process'? To attempt to answer this question the
following experiment was performed. Since scallop EDTA light chains
do not contain thiol groups (Fig. 3), regulatory light chains from the
clam, *Spisula*, which contain thiol groups that can be radioactively
labelled by reacting with (^{14}C) iodoacetic acid, were used to facilitate
identification of the light chain released. The 'radioactively' labelled clam light chains were added to desensitized scallop myosin, about
one mole of added light chain bound and calcium regulation was fully
restored. Subsequent treatment of this complex with EDTA led to the
release of one mole of light chain, roughly half scallop EDTA and half
radioactive clam light chain. These results would suggest that the
release of light chains is random. It is possible that when one EDTA
light chain is removed from the double-headed myosin molecule it may
alter the 'myosin structure' in such a way that the release of the
second light chain is prevented.

4. Possible Models to Explain the Loss of Regulation

The loss of regulation as a result of the release of only one light
chain may be explained by a model where calcium regulation requires
cooperativity between the myosin heads (Szent-Györgyi et al., 1973).
This speculation is supported by the evidence that subfragment-1

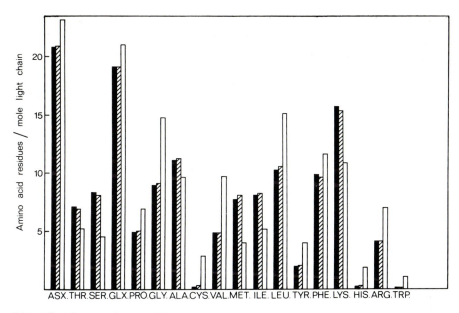

Fig. 3. Comparison of the amino acid analyses of scallop myosin light chains. Solid bars represent the light chain released with EDTA; cross-hatched bars the light chain released from 'desensitized' myosin by treatment with DTNB and open bars the thiol light chains. Note the close agreement between the amino acid composition of the light chains obtained by EDTA and DTNB treatments.
These analyses are the mean values for 24 and 72 h hydrolysates, except that the values of threonine and serine were obtained by extrapolation to zero time hydrolysis, and those of valine and isoleucine were taken from the 72 h hydrolysates. Cysteine values were obtained as cysteic acid from the 24 h hydrolysates of performic acid-oxidised protein

(single myosin head), although it contains a significant amount of the EDTA light chains (about 0.5 moles/mole S_1) and binds calcium, is not regulated by calcium. On the other hand, heavy meromyosin which contains both heads joined by a short tail, requires calcium for interaction with actin. Since both heavy meromyosin and subfragment-1 are produced as a result of mild proteolytic cleavage of myosin, it was considered unlikely that the loss of regulation in subfragment-1 was caused simply by exposure to the proteolytic enzymes, but this explanation cannot be completely excluded. Although these results support a model where regulation requires cooperative interactions between the two myosin heads, there is no direct evidence that such an interaction occurs.

The loss of regulation as a result of light chain release could be explained if only one head at a time interacts with actin, so that the ATPase activity of desensitized myosin in the absence of calcium is nearly maximal even if only one head is working, i.e. the head without the EDTA light chain. An alternative explanation is that loss of regulation may be due to 'interference' between the two myosin heads. For example, if we assume that during desensitization, one light chain is removed from each myosin molecule, then the loss of regulation may be due to the light chain-free myosin head physically interfering with the head that still contains light chain in such a way that the ATPase activity remains high in the absence of calcium. To probe for a possible

modification of the 'myosin structure' which might support these ideas,
double reciprocal plots of the actin activated Mg^{2+}-dependent ATPase
of desensitized and intact myosin were compared. The values obtained
for the maximum turnover rate (ν max) for desensitized myosin (3.8 s^{-1})
were consistently lower than those obtained for intact myosin (6.3 s^{-1})
while the dissociation constants (Km) remained about the same. Similarly heavy meromyosin obtained from desensitized myosin always gave significantly lower values for ν max for its actin-activated ATPase when
compared with that of intact heavy meromyosin, although the actual
values obtained varied between preparations. The extreme susceptibility
of scallop myosin to tryptic digestion (Bárány and Bárány, 1966) makes
it extremely difficult to prepare homogeneous heavy meromyosin preparations. The maximum turnover rates (ν max) for the actin-activated
ATPase of subfragment-1 preparations from desensitized and intact myosins, however, showed no appreciable differences. It is tempting to
suggest that the decrease in catalytic activity and loss of regulation
may either be due to 'interaction' between heads in the same myosin
molecule or the result of one myosin head being permanently 'switched
off'. Experiments are now in progress which may allow us to decide between these various speculations. We can exclude any significant head
to head aggregation between myosin molecules, since desensitized and
native myosins give identical sedimentation profiles in the analytical
ultracentrifuge. If the structure of myosin is altered during 'desensitization', then the process appears to be freely reversible in the
presence of added light chains. The light chains readily recombine with
desensitized myosin, restoring calcium regulation, and the actin-activated ATPase activity (ν max 5.6 s^{-1}) is close to that of the original
myosin.

III. The Mechanism of Myosin-linked Regulation

1. Comparative Aspects of Thin Filament and Myosin Regulation

The calcium-regulating mechanism in vertebrate skeletal muscle, involving the troponin-tropomyosin complex on the thin filament, has
been studied extensively (see review by Weber and Murray, 1973) but
little is known about the mechanism of myosin-linked regulation. Both
calcium regulatory systems work by 'switching off' the muscle at the
same concentration of free calcium, i.e. below 1 μM. In the actomyosin
ATPase cycle, ATP is bound to myosin and is rapidly hydrolyzed to form
the myosin. ADP.P* intermediate complex (Lymn and Taylor, 1970). The
release of this complex is the rate-limiting step in the cycle, the
rate of which is accelerated by combination with actin (Lymn and Taylor, 1971). Recent nucleotide binding experiments by Marston and Lehman (1974) have shown that in vertebrate and molluscan muscles, calcium
controls the same step in the ATPase cycle, the release of ADP.Pi from
the myosin ADP.Pi* steady-state intermediate complex, regardless of
whether regulation is on the thin filament or on the myosin. However
there is an apparent difference between the two systems. In the actin-
linked system, high ATP concentrations are required to prevent the
formation of 'rigor links' between nucleotide-free myosin and actin,
which are capable of 'activating' the actin filaments, even in the absence of calcium, and allowing contraction to occur (Bremel and Weber,
1972). In molluscan muscles, even at low ATP concentrations, contraction does not occur in the absence of calcium. However the presence of
nucleotide bound to the myosin, presumably as the steady-state intermediate (myosin ADP.Pi*) appears to be required for myosin-linked regulation, since in molluscan muscles, 'rigor linkages' are formed between nucleotide-free myosin and actin even in the absence of calcium.

2. Possible Models of Myosin-linked Regulation

In a scallop myosin molecule our evidence indicates that each head contains an EDTA light chain. Therefore a simple model of regulation could be envisaged where the light chain is bound near the actin combining site on the myosin head so that it is capable of "turning off" this site in the absence of calcium by sterically blocking interaction with actin. The mechanism could involve either calcium-induced conformational changes within the light chain or in the whole myosin molecule. However, the exact position of the light chain on the myosin head is unknown.

Despite our almost complete ignorance about the location of the light chain, can we speculate whether each light chain is attached to the myosin heavy chain by a single site or by more than one? There is evidence that loss of calcium regulation can be achieved without release of the EDTA light chain, for example, mild heat treatment (45° for 5 min) modifies the myosin in such a way that calcium regulation is completely abolished without release of the light chain (Szent-Györgyi et al., 1975). Similarly, preliminary results with the sulphydryl reagent, N-ethylmaleimide, also indicate a loss of regulation without a detectable release of the light chain (Kendrick-Jones and Jakes, unpublished observations). Further evidence is provided by the rabbit DTNB light chain which always binds to desensitized scallop myosin preparations, but frequently there is no restoration of calcium regulation (Kendrick-Jones, 1974).

These results could be explained if we assume there are two types of light chain binding sites on the myosin. One site, so that the light chain always remains attached to the myosin in the presence of divalent cations and a second site, to which the light chain must be able to attach in order to operate the calcium switch. Then, in the absence of calcium, when the light chain is attached to this second site, it may directly block myosin interaction with actin. The increase in calcium concentration on activation of the muscle could lead to calcium binding to the light chain or to the whole 'head' and the resulting 'conformational' change might break the attachment at this second site, 'switching on' the myosin so that it may interact with actin. A similar mechanism has previously been proposed for the regulation by troponin and tropomyosin on the thin filament in vertebrate skeletal muscles (Margossian and Cohen, 1973; Hitchcock et al., 1973; Drabikowski et al., 1974). Experiments are now in progress to test the validity of this model for myosin-linked regulation.

3. Distribution of the Calcium Control Systems

Comparative studies on the evolution of the regulatory systems have established that some species have actin-linked regulation, some myosin-linked, and some have both (Lehman et al., 1972). Almost all vertebrate muscles, such as the skeletal muscles from rabbit, chicken, frog and mouse and cardiac muscle from dog are controlled via troponin on the thin filament (Ebashi et al., 1966; Lehman et al., 1972) (Table 2). It was previously thought that myosin-linked regulation was restricted to 'primitive' invertebrate muscles, since it was only found in nemertine worms (*Cerebratulus*), the sea cucumber (*Thyone*), the brachiopod (*Glottidia*) and in a variety of molluscs such as gastropods, lamellibranchs (scallops and clams) and cephalopods (squid). Recently, however, using a simple test devised by Lehman and his colleagues to determine which regulatory system is present in a muscle, Bremel (1974) has shown that vertebrate smooth muscle also contains a myosin-linked regulatory system. In the more 'sophisticated' invertebrates, such as

Table 2. Distribution of calcium control systems in different species

	Species	Muscle type	Thin filament regulation - troponin	Myosin regulation - light chains
Vertebrate	Rabbit	striated	+	-
"	Chicken	smooth (gizzard)	-[c]	+[a]
Arthropod	Lobster	striated	+	-
	Insect	striated	+	+[b]
Mollusc	Scallop	striated	-	+
	"	smooth	-	+
Annelids	Earth-worm	striated	+	+

[a] Bremel (1974).

[b] Lehman et al. (1974).

[c] There is preliminary evidence (S.V. Perry, personal communication) that vertebrate smooth muscle contains troponin components on the thin filaments. The remaining data taken from Lehman et al. (1972).

the annelid worms (earthworm and marine worms) and insects, both troponin and myosin-linked regulatory systems are present in the same muscle (Lehman et al., 1972, 1974). These invertebrates species occupy a position midway between molluscs and vertebrates on the evolutionary tree and it was thought that they might represent the crossover point from 'primitive' myosin-linked regulation to the more elaborate troponin system. The recent observations that vertebrate smooth muscle contains myosin-linked regulation and that troponin exists in invertebrate species, such as nematodes, which preceded molluscs in evolution (Lehman and Szent-Györgyi, personal communication) would suggest that the two regulatory systems evolved on parallel pathways, and that myosin-linked regulation has been retained during the evolution of higher vertebrates. It is therefore conceivable that all muscles may have, or may have had, both types of calcium regulatory systems. This speculation is further supported by the existence of light chains in vertebrate myosins which may play a role in calcium control (Kendrick-Jones, 1974) and the observation that even in molluscan muscles, there are small amounts of 'troponin-like' components on the thin filaments (Kendrick-Jones, unpublished observation).

The two regulatory systems appear to be functionally unrelated; myosin light chains do not replace any of the troponin components and form functional units and similarly the troponin components do not replace the molluscan light chains and regulate myosin (Hitchcock and Kendrick-Jones, 1975). However, there appears to be extensive sequence homology between the calcium-binding component of troponin (TN-C) and the rabbit myosin 'alkali' light chains (Collins, 1974; Weeds and McLachlan, 1974) which suggests a common ancestral gene. Studies are now in progress on the sequence of the scallop EDTA light chain to determine whether it is also 'structurally related' to troponin-C.

The reasons for the existence of these two distinct calcium regulatory systems, their evolutionary significance and what physiological advantage is conferred by the possession of actin-linked, or myosin-linked or both systems, are interesting questions which remain to be answered.

4. The Possible Existence of Myosin Control in Vertebrate Skeletal Muscles

Although there is extensive evidence that vertebrate skeletal muscles contain thin filament regulation (see review by Ebashi et al., 1969), there are a number of observations which suggest that some 'type' of calcium control on the myosin may also exist. There are advantages in regulating both the myosin and the thin filament. In insect muscle, both systems act independently and increase the sensitivity of the contractile apparatus to calcium, thus ensuring that the steady-state rate of ATP hydrolysis in relaxed muscle is minimal (Lehman et al., 1974).

Huxley and Brown (1967), in their detailed study of the X-ray diffraction patterns of live muscle, observed that the myosin cross-bridges (heads) move when the muscle contracts. Haselgrove (1970, 1975) and later Huxley (1972), using muscles which had been stretched to lengths where there should be either reduced overlap or no apparent overlap between the myosin and actin filaments, further showed that the movement of the cross-bridges as a result of activation was not due directly to interaction with actin. The simplest interpretation of these results is that the myosin cross-bridges are able to respond directly to the calcium released as a result of activation. However, Huxley (1972) has expressed caution in this interpretation, since the possibility cannot be excluded that some overlap of the filaments may still exist in local areas of the fibre, and the changes in these overlap regions as a result of activation may be transmitted in a cooperative manner along the remainder of the myosin filaments.

Evidence for structural changes in vertebrate thick filaments induced by calcium in the range required for activation of living muscle has recently been obtained by Morimoto and Harrington (1974). They found an increase in the sedimentation coefficient and a decrease in the relative viscosity of native thick filaments with increasing calcium concentration (from 10^{-7} to 10^{-5} M) which may be explained by some structural change in the filament induced by the binding of calcium. They also showed that vertebrate myosin has two high affinity calcium binding sites located on the 'DTNB' light chain. In a detailed study of calcium binding to rabbit skeletal myosin, Bremel and Weber (1975) observed that at 20 μM calcium and a magnesium concentration of 100 μM, a myosin molecule contains two high-affinity calcium binding sites. Magnesium alters calcium binding in a complex manner, not by simple competition. However, even at a free magnesium concentration of 1 mM, each myosin molecule binds one calcium at a free calcium ion concentration (20 μM) where the muscle is fully activated.

It is difficult to visualise at present how calcium binding to the DTNB light chains could produce the large-scale movements of the myosin cross-bridges observed in the X-ray diffraction studies. It has been proposed (Werber et al., 1972) that if the light chains are associated with the 'hinge' connecting the head with the myosin rod, then calcium binding to the light chains could produce conformational changes in this hinge region which might influence the movement of the heads.

The X-ray diffraction, hydrodynamic and calcium binding studies provide evidence for a calcium-sensitive system on vertebrate myosin. Whether such a system is analogous to that in molluscan myosins remains to be established. Rabbit skeletal myosin contains two classes of light chains: (1) the 'alkali' light chains (Fig. 6) which are thought to be required for the enzymic activity of myosin (Stracher, 1969) and may determine the maximum shortening velocity of the muscle (Weeds et al., 1974); and (2) a light chain with a molecular weight of about

19,000 referred to as the DTNB light chain, since it may be partially removed from myosin by treatment with 5,5'-dithiobis-(2-nitrobenzoic acid) (DTNB) (Weeds, 1969; Gazith et al., 1970). The function of this light chain is unclear as there is no direct evidence that it is required for the normal biochemical functions of the isolated myosin (Weeds and Lowey, 1971). However, there is now evidence that it binds calcium (Werber et al., 1972) and can be phosphorylated by a specific kinase (Perrie et al., 1973) which might also suggest that it has a regulatory role. This speculation is supported by the observation that this rabbit DTNB light chain will bind to 'desensitized' scallop myosin and restore calcium regulation, i.e. it will functionally replace the scallop 'EDTA' light chain (Fig. 4) (Kendrick-Jones, 1974). The light

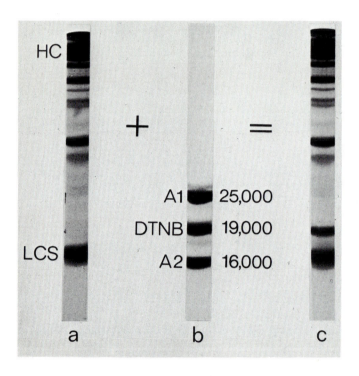

Fig. 4a-c. Specificity of light chain interaction with 'desensitized' scallop myosin. 10% SDS-acrylamide gels of (a) initial desensitized scallop myosin, (b) rabbit light chains (released from rabbit myosin by treatment with 4 M urea (Weeds and Lowey, 1971), (c) desensitized scallop myosin after mixing with rabbit light chains and extensive washing to remove the unbound light chains. Note, only the DTNB light chain (19,000) binds to the myosin. HC, myosin heavy chains; LCS, scallop light chains; A1 and A2, rabbit alkali light chains; DTNB, 'DTNB' light chain

chain attachment sites on scallop myosin are specific for the EDTA and rabbit DTNB light chains (and related light chains); the other rabbit light chains, the alkali light chains, do not bind (Fig. 4). Both EDTA and DTNB light chains are bound to scallop myosin with about the same high affinity, and comparable amounts of each light chain are required to restore calcium regulation to desensitized preparations (Fig. 5).

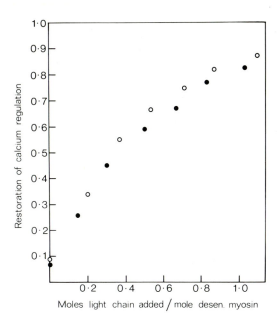

Fig. 5. Restoration of calcium regulation on addition of scallop EDTA light chain (●) or rabbit DTNB light chain (o) to desensitized scallop myosin. Desensitized myofibrils were mixed with increasing amounts of the light chains (assuming a 65% myosin content and molecular weights for myosin, 450,000; EDTA light chain, 16,500 and DTNB light chain, 18,500) and the actin-activated Mg^{2+} ATPase activity measured (under the conditions outlined in Table 1) in the absence of calcium and compared with that in the presence of calcium, i.e. restoration of calcium regulation = (1 - ATPase at $\dfrac{10^{-8} \ M \ Ca^{2+}}{10^{-4} \ M \ Ca^{2+}}$). Almost complete recovery of calcium regulation is achieved when about one mole of either light chain is added

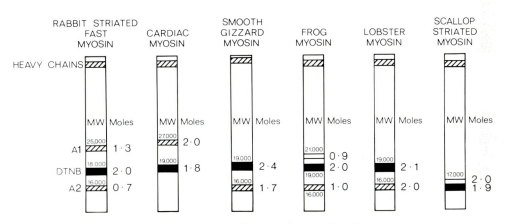

Fig. 6. Diagrammatic representation of the subunit structure of various myosins. The light chain compositions (moles light chain myosin[-1]) of the different myosins were established by densitometry of acrylamide gels and verified, where possible, by a [14C] iodoacetic acid labelling technique (Kendrick-Jones et al., 1975). Note each myosin contains about two moles of light chain with a molecular weight (MW) of about 19,000. These light chains will 'functionally replace' the scallop EDTA light chain in molluscan myosin

All the myosins we have studied from a number of vertebrate and invertebrate species and a variety of different muscle types contain about two moles of a light chain with a molecular weight of about 19,000 which will specifically bind to desensitized scallop myosin and restore regulation (Fig. 6). The recent demonstration of myosin-linked regulation in vertebrate smooth muscle, and this demonstration that its myosin contains a light chain which will replace the scallop regulatory light chain, would further support the idea that the 19,000

134

light chains in all these myosins have a regulatory function. It is difficult to believe that these light chains are not involved in some type of calcium control in their parent myosins, since one would have expected that during the millions of years of evolution that separate vertebrates from molluscs, mutation and natural selection would have removed such a non-functional light chain.

Although there is clear evidence that vertebrate skeletal myosins bind calcium, and are able to respond to changes in the free calcium ion concentration in the range where muscle is activated (from 10^{-7} to 10^{-5} M) and contain the necessary regulatory light chains, there is no direct demonstration that the calcium bound to the myosin is involved in controlling the interaction of myosin and actin *in vitro*. In fact there is evidence that calcium binding to rabbit myosin actually inhibits its actin-activated ATPase in the presence of unregulated actin (Bremel and Weber, 1975). However, it is questionable how far *in vitro* studies are relevent when, *in vivo*, the myosin and actin filaments are in highly-organised lattice arrays and are under tension, which may well alter the relative 'interaction' of these proteins involved in the contractile cycle. Also it is conceivable that during isolation and purification of the myosin, calcium regulation could have been irreversibly lost. For example, insect myosin, which is calcium-regulated in the intact muscle, easily loses this function when the myosin is isolated from the muscle (Lehman et al., 1974). Obviously the final answer to this fascinating question of whether there is a calcium 'switch' on the thick filaments in vertebrate skeletal muscle requires detailed biochemical studies under conditions where the proteins are as close as possible to their *in vivo* configuration.

References

Bárány, M., Bárány, K.: Biochem. Z. 345, 37-56 (1966).
Bremel, R.D.: Nature 252, 405-407 (1974).
Bremel, R.D., Weber, A.: Nature 238, 97-101 (1972).
Bremel, R.D., Weber, A.: Biochim. Biophys. Acta 376, 366-374 (1975).
Collins, J.H.: Biochem. Biophys. Res. Commun. 58, 301-308 (1974).
Drabikowski, W., Barylko, B., Dabrowska, R., Nowak, E., Szpacenko, A.: In: Calcium Binding Proteins, pp. 69-107. Warszawa: PWN-Polish Scientific Publishers, and Amsterdam: Elsevier 1974.
Ebashi, S., Endo, M.: Prog. Biophys. Mol. Biol. 18, 123-183 (1968).
Ebashi, S., Endo, M., Ohtsuki, I.: Quant. Rev. Biophys. 2, 351-384 (1969).
Ebashi, S., Iwakura, H., Nakajima, H., Nakamura, R., OOI, Y.: Biochem. Z. 345, 201-211 (1966).
Gazith, J., Himmelfarb, S., Harrington, W.F.: J. Biol. Chem. 245, 15-22 (1970).
Greaser, M.L., Yamaguchi, M., Brekke, C., Gergely, J.: Cold Spring Harbor Symp. Quant. Biol. 37, 235-244 (1972).
Haselgrove, J.C.: X-ray Diffraction Studies on Muscle. Ph.D. Thesis, University of Cambridge, 1970.
Haselgrove, J.C.: J. Mol. Biol. 92, 113-143 (1975).
Hitchcock, S.E., Huxley, H.E., Szent-Györgyi, A.G.: J. Mol. Biol. 80, 825-836 (1973).
Hitchcock, S.E., Kendrick-Jones, J.: In: Calcium Transport in Contraction and Secretion. Symp. Bressanone, Italy. Amsterdam: Elsevier 1975.
Huxley, H.E.: Cold Spring Harbor Symp. Quant. Biol. 37, 361-376 (1972).
Huxley, H.E., Brown, W.: J. Mol. Biol. 30, 385-434 (1967).
Kendrick-Jones, J.: Nature 249, 631-634 (1974).

Kendrick-Jones, J., Lehman, W., Szent-Györgyi, A.G.: J. Mol. Biol. 54, 313-326 (1970).
Kendrick-Jones, J., Szentkiralyi, E.M., Szent-Györgyi, A.G.: Cold Spring Harbor Symp. Quant. Biol. 37, 47 (1972).
Kendrick-Jones, J. Szentkiralyi, E.M., Szent-Györgyi, A.G.: The regulatory role of the myosin light chains. In preparation (1975).
Lehman, W., Bullard, B., Hammond, K.: J. Gen. Physiol. 63, 553-563 (1974).
Lehman, W., Kendrick-Jones, J., Szent-Györgyi, A.G.: Cold Spring Harbor Symp. Quant. Biol. 37, 319-330 (1972).
Lymn, R.W., Taylor, E.W.: Biochemistry 9 (1970).
Lymn, R.W., Taylor, E.W.: Biochemistry 10, 4617-4624 (1971).
Margossian, S.S., Cohen, C.: J. Mol. Biol. 81, 409-413 (1973).
Marston, S., Lehman, W.: Nature 252, 38-39 (1974).
Morimoto, K., Harrington, W.F.: J. Mol. Biol. 88, 693-709 (1974).
Perrie, W.T., Smillie, L.B., Perry, S.V.: Biochem. J. 135, 151-164 (1973).
Stracher, A.: Biochem. Biophys. Res. Commun. 35, 519-525 (1969).
Szent-Györgyi, A.G., Szentkiralyi, E.M., Kendrick-Jones, J.: J. Mol. Biol. 74, 179-203 (1973).
Weber, A., Murray, J.M.: Physiol. Reviews 53, 612-673 (1973).
Weeds, A.G.: Nature 223, 1362-1364 (1969).
Weeds, A.G., Lowey, S.: J. Mol. Biol. 61, 701-725 (1971).
Weeds, A.G., McLachlan, A.D.: Nature 252, 646-649 (1974).
Weeds, A.G., Trentham, D.R., Kean, C.J.C., Buller, A.J.: Nature 247, 135-139 (1974).
Werber, M., Gaffin, S., Oplatka, A.: J. Mechan. Cell Motility 1, 91-96 (1972).

Discussion

Dr. Weber: Removal of regulatory light chain gave you a 50% reduction in the actin-activated ATPase activity; were you able to restore the actin-activated ATPase activity by readding the regulatory light chain?

Dr. Kendrick-Jones: Yes, there is a partial restoration of ATPase activity on addition of the regulatory light chain. The data from different myosin preparations are rather variable, but in all cases there is a significant increase in the actin-activated ATPase activity of the desensitized myosin when the regulatory light chain is recombined. The decrease in activity does not appear to be due to aggregation between heads of different myosin molecules, since the sedimentation profiles of native and desensitized myosins are very similar. We suspect that the loss of activity may be due to some type of 'interaction' between the heads of the same myosin molecule caused by the 'desensitization procedure'.

Dr. Weber: We find the same ATPase activity per head for HMM and S_1, although S_1 no longer has the regulatory light chain (according to its loss of Ca binding) and HMM has the light chain.

Dr. Gergely: Is it correct that even when you remove divalent cations only one mol of light chain is removed by DTNB per heavy chain?

Dr. Kendrick-Jones: Yes, that is correct; DTNB selectively removes only the regulatory light chain.

Dr. Gergely: Returning to Dr. Weber's remark, what is the evidence for the absence of LC2 in S_1, since nicks in the light chain could make it look in SDS gels as though it wasn't there?

Dr. Kendrick-Jones: That is certainly a problem when using SDS gels for diagnosis. The light chain could possibly still be present on the S_1, but maybe 'nicked', and dissociation in SDS could release the light chain as small fragments which may not be observed on the gels.

Dr. Perry: If you take S_1 from rabbit myosin and try to phosphorylate it with myosin light chain kinase you don't get any phosphorylation; therefore a fully intact S_1 with that site is not present.

Dr. Weber: Our criterium for the loss of light chain is not based on SDS gels; our criterium is a complete loss of calcium binding ability.

Dr. Perry: I wonder if you have any ideas on why during evolution Ca regulation moved from the A to the I filament?
Myosin regulation seems to be the most primitive form and regulation through the thin filament system would appear to be correlated with specialization and an increase in the speed of contraction. It is not easy to understand why regulation through the filament is more appro-ciate for fast muscle than A filament regulation where Ca acts directly on the myosin ATPase. The regulatory system of the thin filament has also the disadvantage of being associated with every seventh actin monomer, it thereby being necessary to convey the effect of Ca to the other six actin monomers.

Concerted Regulation of Glycogen Metabolism and Muscle Contraction

E. H. Fischer, J.-U. Becker[1] H. E. Blum[1] B. Byers, C. Heizmann[2]
G. W. Kerrick, P. Lehky, D. A. Malencik, and S. Pocinwong

Whenever a muscle contracts, glycogen immediately breaks down to provide the energy necessary to maintain contraction. We propose to discuss here how these two physiological processes can be simultaneously
triggered and regulated. Also, since we have been wondering how such
very complex reactions originally arose and evolved with time, we have
been looking at certain evolutionary aspects of these control mechanisms.

A. Regulation of Carbohydrate Metabolism

The regulation of carbohydrate metabolism seems particularly appropriate for such a study (Fischer et al., 1970). First, the reactions that
bring about the synthesis and breakdown of glycogen are essentially
ubiquitous in all forms of life, from the simplest prokaryotic cell
to the complex tissues of higher plants and animals. And since life
is believed to have emerged on earth in a reducing atmosphere, it is
quite probable that the anaerobic pathways of metabolism preceded
aerobic metabolism by several hundred million years. Second, there
are two main ways in which metabolic processes can be regulated: by
allosteric mechanisms and by the covalent modification of enzymes.
Both glycogen synthase and phosphorylase, the two enzymes most directly involved in the synthesis and breakdown of glycogen, are controlled
by allosteric and covalent modifications. Third, the covalent control
of phosphorylase is not catalyzed by a single enzyme, but by a number
of enzymes acting successively on one another. Furthermore, this cascade of enzymatic reactions is closely linked to other physiological
processes such as hormone stimulation and the nerve impulse that triggers contraction. Finally, one can expect that the enzymes involved
will contain a multiplicity of functional sites to account for these
multiple interactions. In turn, these sites can be used as evolutionary
markers to define how complex regulatory proteins were originally
constructed, then modified in the course of time.

I. Allosteric vs. Covalent Control Mechanisms and Cascade Reactions

Before proceeding, there are two basic questions to consider: First,
why did organisms find it necessary to regulate the activity of certain
key enzymes by both allosteric and covalent mechanisms? In either instance, the inactive form of an enzyme is converted into an active

---

[1] Recipient of fellowships from the Studienstiftung des Deutschen
Volkes and the Deutsche Forschungsgemeinschaft.

[2] Recipient of a Swiss National Foundation Research Fellowship.

Allosteric Control

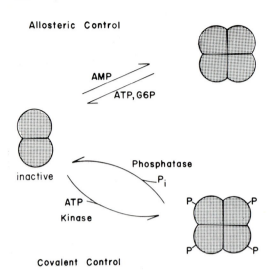

AMP

ATP, G6P

inactive

Phosphatase

P_i

ATP

Kinase

P

P

P

P

Covalent Control

Fig. 1. Allosteric and covalent control of muscle phosphorylase

species (Fig. 1). The answer is this (Fischer et al., 1971): if one had to rely solely on allosteric activation, unless strict intracellular compartmentation existed one would have to activate or inhibit all the enzymes susceptible to the same allosteric effectors. In the case of phosphorylase, the enzyme is activated by AMP, inhibited by ATP, G6P etc. But many enzymes of glycolysis or of the Krebs cycle are also affected by AMP, ATP or G6P. In other words, one would have to open simultaneously all these doors or close many others. By contrast, since covalent control is enzyme-catalyzed and the regulatory enzymes involved are usually highly specific, it provides the possibility of affecting a single step, i.e. of opening or closing a single door, without necessarily having to touch any other. This would seem to be mandatory if one wanted to build up the concentration of certain metabolites or, perhaps, divert their flow from one pathway to another. Furthermore, covalent modification "freezes" an enzyme in a given conformation, thereby protecting it from modulation by the usual host of allosteric effectors.

Secondly: why do we need several regulatory enzymes acting successively on one another to ultimately activate or inhibit a single target enzyme (Fig. 2)? An obvious explanation is that it provides for a considerable physiological amplification. Table 1 lists some molecular and enzymatic properties for the four enzymes involved in glycogen breakdown. There are two or three ways in which the approximate level of amplification provided by this cascade system can be measured; most are poor because of the uncertainty in the activity of the membrane-bound adenylate cyclase, and the rate of the back reactions. Nevertheless, they all yield an amplification of the order of a million-fold. This value is consistent with that obtained experimentally. During contraction, approximately 20 μmol of lactic acid can be produced per min. per ml of intracellular water, corresponding to the degradation of 10 mM glucosyl residues per min. Assuming that the concentration of epinephrine is of the order of 10^{-8} to 10^{-9} M, this would also indicate a signal amplification of approximately one million-fold. This amplification is not necessarily one of rate, since one is always limited by the slowest reaction, but certainly in the amount of material that can be mobilized under the influence of extremely minute concentrations of a compound serving as a metabolic signal. This is why trace amounts of a hormone can result in the mobilization of large quantities of glycogen within

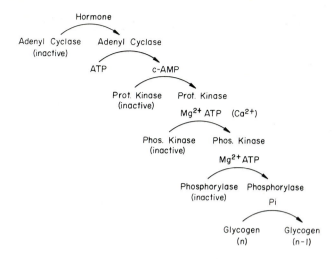

Fig. 2. Cascade reaction in the control of glycogen breakdown. For reasons of simplicity, the reverse processes are omitted

Table 1. Structure and activity of enzymes involved in glycogen metabolism

Enzyme	Subunit structure	MW (catalytic subunit is italicized	Concentration[a]		Specific activity (μmole substrate converted/mg/min)
			mg/ml	μM	
Adenyl cyclase	unknown	158,000 *38,000*	particulate		1-5[b]
Protein kinase	$\alpha_2\beta_2$	α 82,000 β *49,000*	0.0065	0.23	2
Phosphorylase[c] kinase	$(\alpha\beta\gamma)_4$	α 131,000 β *118,000* γ 45,000	0.84	4	30[d]
Phosphorylase	α_2 or α_4	α 100,00	4.2	100	30[e]

[a] Assuming intracellular water at 75% of tissue wet weight.

[b] Estimated value from impure preparations.

[c] From dogfish skeletal muscle.

[d] At pH 8.2, and calculated on the basis of phosphate group transferred per phosphorylase subunit.

[e] Measured in the direction of glycogen breakdown.

a few seconds, so that amplification is certainly a factor, but probably not the most important one.

The most important feature of cascade reactions is that each of the "converter" or "regulatory" enzymes allows the linkage of one metabolic or physiological process to another (Fischer et al., 1972). Considering one more the regulation of carbohydrate metabolism (Fig. 3), the ade-

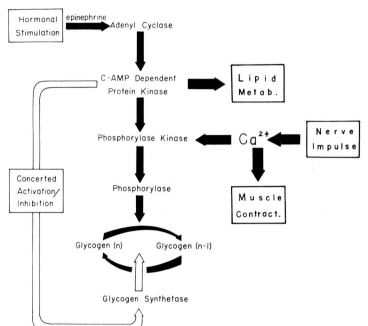

Fig. 3. Linking of metabolic pathways through control enzymes

nylate cyclase system can be viewed as linking carbohydrate metabolism to hormonal stimulation. Phosphorylase kinase links this process to muscle contraction: upon nerve impulse, calcium released from the sarcoplasmic reticulum triggers contraction by binding to troponin, and glycogen breakdown by binding to phosphorylase kinase. Even the third enzyme involved in this regulatory cascade, the c-AMP-dependent protein kinase, can link carbohydrate metabolism to lipid metabolism by activating the hormone-sensitive lipase (Krebs, 1972; Huttunen et al., 1970). Also, by acting on glycogen synthase (Larner and Villar-Palasi, 1971), it will prevent a futile cycle engendered by the simultaneous synthesis and breakdown of glycogen. Indeed, while phosphorylation <u>activates</u> phosphorylase, it <u>inhibits</u> glycogen synthetase; this synchronous control of both enzymes provides for a safeguard against the useless utilization of ATP.

There is a last reason for having multiple enzymes acting on one another to ultimately affect a single metabolic step: it would be totally impossible to store all the information needed for concerted regulation within a single protein molecule. Consider phosphorylase, for instance. The enzyme must contain an active site which binds each of its three substrates, glycogen, P_i and G1P (Fischer et al., 1970). There is an allosteric site which binds AMP or ATP (they compete with one another) and separate sites for the binding of G6P, perhaps UDPG, etc., which also affect the activity of the enzyme. There is the site of covalent control, that is, the region that becomes phosphorylated in the phosphorylase b to a conversion and the site binding pyridoxal phosphate, even though we do not know precisely how this cofactor functions in the catalytic process. Finally, there are at least two regions involved in the aggregation of the enzyme and accounting for the formation of phosphorylase b dimer and phosphorylase a tetramer. Whether or not some of these sites overlap is not know, but it is obvious that a considerable portion of the real estate of the enzyme must be devoted to these functional sites. It would be quite inconceivalbe that we could further introduce, in the same molecule, all the structural features required for the binding of calcium that links carbohydrate metabolism to muscle contraction, or the receptors allowing for hormonal recognition. One is left with no alternative but to store some of the additional information in "annex" molecules, as one would store excess memory in an "annex" computer when the first one has become saturated.

The same situation prevails in the regulation of *E. coli* glutamine synthetase (Adler et al., 1973): the enzyme can be modified by adenylylation; this reaction is catalyzed by another enzyme, itself modified by a third protein that can be uryldylylated. Each of these separate regulatory enzymes can now be modulated by different effectors such as ADP, P_i, α-ketoglutarate, etc. Here again, it would be quite impossible to package all the information needed into a single protein molecule, all the more since the target enzyme itself, glutamine synthetase, is cumulatively inhibited by 7 or 8 end-products of glutamine metabolism: histidine, tryptophan, alanine, CTP, AMP etc. In fact, this is certainly the reason why some of the regulatory enzymes themselves are made up of different subunits - as will be detailed later for phosphorylase kinase: all have separate functions and were brought together in the course of evolution to fulfill a complex concerted task.

B. The Site of Covalent Control in Glycogen Phosphorylase

The site involved in the covalent control of phosphorylase is located at the N-terminus of the molecule (Titani et al., 1975): the single residue that is phosphorylated occupies position number 14 in a peptide chain containing ca. 865 residues (Fig. 4). There is nothing outstanding about this structure: the phosphoseryl residue is sandwiched be-

```
        1                       5                              10
N-Acet · Ser · Arg · Pro · Leu · Ser · Asp · Gln · Glu · Lys · Arg ·
       11                      P    15                         20
      · Lys · Gln · Ile · Ser · Val · Arg · Gly · Leu · Ala · Gly ·
       21                      25
      · Val · Glu · Asn · Val · Thr · Glu · Leu · Lys · Lys
```

Fig. 4. Sequence of the phosphorylated site in rabbit muscle phosphorylase. (From Titani et al., 1975)

tween two hydrophobic amino acids, followed by an arginine that appears to play a crucial role in the specificity of phosphorylase phosphatase. This enzyme will not act on a number of synthetic phosphoseryl peptides, but will dephosphorylate phosphopeptides obtained from phosphorylase as long as the arginyl residue is present (Graves et al., 1960). Then, on the distal side, one notices the regular appearance of hydrophobic residues every three amino acids: Val-x-x-Leu-x-x-Val-x-x-Val-x-x-Leu..

(Titani et al., 1975). This might indicate the presence of an α-helix in which every third residue is directed towards the hydrophobic interior of the molecule, as shown for some of the α-helices in hemoglobin and myoglobin (Perutz et al., 1965).

Why would the site of covalent control be at the N-terminus and would this be expected for other enzymes subjected to covalent regulation? There is no answer as yet, since this is the only case where this kind of information is available. True, if one looks at the conversion of trypsinogen to trypsin, one sees that cleavage of a single peptide bond at the N-terminus results in the activation of the enzyme (Walsh, 1970); but then, of course, this is a totally different kind of reaction and, perhaps, a mere coincidence.

The phosphorylated site must accommodate the two enzymes responsible for its modification (phosphorylase kinase and phosphatase) and therefore must be exposed and perhaps somewhat flexible; furthermore, since phosphorylation leads to tetramerization, some of the residues might be directly implicated in subunit interaction. Perhaps one is dealing with a situation somewhat similar to that observed with lactic dehydrogenase (Adams et al., 1970). In this instance, the first 22 residues from the N-terminus form an extended arm detached from the rest of the molecule. This segment plays an important role in subunit interaction, particularly residues 7 to 10, located close to a single turn of an α-helix. Whether this applies to phosphorylase will have to await the X-ray crystallographic data of Louise Johnson in David Philip's laboratory in Oxford.

The site of covalent control in phosphorylase has been highly conserved throughout evolution, at least since the early vertebrates. By contrast, if one considers the phosphorylated site in yeast phosphorylase, one finds that it is totally different (Lerch et al., 1975). In this instance, a threonyl rather than a seryl residue is phosphorylated, and the overall region of the molecule is acidic rather than highly basic. This is not surprising considering that *Saccharomyces* separated from the main line of evolution some 1.5 billion years ago and, of course, carbohydrate metabolism in yeast would be expected to be controlled by a totally different set of cellular or nutritional requirements. On the other hand, the site involved in the binding of pyridoxal phosphate, the coenzyme of phosphorylase, has been extraordinarily conserved; here, 14 out of 17 residues are identical (Lerch et al., 1975). This is perhaps the strongest evidence yet, though still indirect, that pyridoxal-P is directly involved in the catalytic reaction. Otherwise, why would the structure of this segment of the molecule be retained essentially unchanged for more than a thousand million years?

C. Control of Phosphorylase Activity in Relation to Muscle Contraction

As indicated above, glycogen breakdown cannot be initiated unless calcium ions are made available. This is most easily demonstrated in glycogen particles (Meyer at al., 1970) that can be isolated from mammalian muscle and that contain most of the enzymes involved in the regulation of glycogen metabolism: phosphorylase, phosphorylase kinase and phosphatase, glycogen synthase, a debranching enzyme, and many other glycolytic enzymes as well as elements of the sarcoplasmic reticulum. If Mg and ATP are added to this preparation, nothing happens. By contrast, if Ca, Mg and ATP are added, there is an immediate activation (Heilmeyer et al., 1970) of phosphorylase which returns to zero after a few minutes when all the ATP has been consumed; this reaction was termed

"flash activation" of phosphorylase. When calcium is added in excess,
re-addition of ATP alone initiates a second flash activation and the
entire process can be repeated several times. Addition of EDTA or EGTA
greatly reduced or abolished the flash activation, but the process can
be restored by calcium ions.

D. Properties and Subunit Structure of Phosphorylase Kinase

The above effects are all mediated by phosphorylase kinase which has
an absolute requirement for calcium. Therefore, the structure of this
enzyme was studied in some detail. Also, since much of the earlier
work had been carried out on mammalian phosphorylase kinase (Hayakawa
et al., 1973a, 1973b), we thought that it might be of interest to look
at the structure of an enzyme isolated from a more primitive species
with the hope that some of its basic features would not have been ob-
scured by various complexities added in the course of evolution. The
Pacific dogfish (*Squalus acanthias*) was selected because it is a very
primitive vertebrate that separated from the main line of evolution
some 450 million years ago; it possesses a well-developed endocrine
system and is readily available in the Puget Sound area. Furthermore,
the skeletal muscle is predominantly white, surrounded by a few layers
of red muscle cells that can be easily separated if necessary. Some
properties of dogfish and rabbit skeletal muscle phosphorylase kinase
are summarized in Table 2. Both enzymes have molecular weights (MWs)

Table 2. Properties of dogfish and rabbit skeletal muscle phosphorylase
kinase

Properties	Phosphorylase kinase Dogfish	Rabbit
Molecular weight	1.3×10^6	1.28×10^6
Specific activity (pH 8.2)	30	37
Optimum pH	8.3	8.2
pH 6.8/8.2 activity ratio for phosphorylated species:	0.5	0.01 0.40
Substrate specificity	% Relative rates	
Phosphorylase b (dogfish or rabbit)	100	100
TN-I " "	0	10
TN-T " "	0	2
Myosin " "	0	1
ATP	100	100
ATP-γ-S	10	10
GTP	0	10

in the order of 1.3 million and are made up of three types of subunits,
α, β and γ, having MWs of ca. 131,000, 118,000 and 45,000 respectively,
with some differences between the two species. A possible role for
each of these subunits will now be discussed.

Rabbit phosphorylase kinase is active above pH 8 but quite inactive
below 7. Activation at low pH can be achieved by addition of calcium;
activity is further increased by phosphorylation of the protein or by
limited proteolysis. Phosphorylation of the enzyme can occur in two
ways: first, the c-AMP dependent protein kinase appears to modify the
β-subunits; then, the α-subunit can also become phosphorylated in a
subsequent autocatalytic reaction (Cohen, 1974; Hayakawa et al., 1973a).
The total number of phosphate groups introduced is not exactly known.
Recently an ingenious mechanism has been proposed by which the phosphorylation of the α-subunit (second site phosphorylation) allows for
the accelerated dephosphorylation of the β-subunit by a specific phosphatase resulting in the inactivation of the enzyme (Cohen et al.,
1973).

Since activation by limited proteolysis is accompanied by a rapid de-
struction of the α- and β-subunits while γ is resistant, it was originally hypothesized that γ might represent the catalytic subunit of
the enzyme. We would like to propose that β really contains the catalytic site of the enzyme, while α represents a regulatory subunit maintaining β in an inactive state; the possible role of the γ-subunit will
then be discussed.

I. The α- and β-Subunits

When different purified preparations of dogfish phosphorylase kinase
are examined by SDS gels, one finds that while the ratio of α to β
remains constant at 1:1, the proportion of γ subunit can vary from
1 to 8 or so. Yet the specific activities of all these preparations
are essentially the same. Admittedly, not much can be concluded from
these observations except, perhaps, that the γ subunit can undergo
aggregation and that it does not seem to represent the catalytic subunit, at least not by itself.

A stronger argument that the catalytic site resides in the β-subunit
results from the limited proteolysis experiment illustrated in Fig. 5.
One first observes a rapid activation of the enzyme, concomitant with
the disappearance of the α-subunit, which happens to be very susceptible to proteolytic attack. This clearly indicates that α had a regulatory function of maintaining β in the inactive form. Then, as β is
degraded, enzyme activity disappears while the γ-subunit remains essentially unchanged. If partially degraded solutions of phosphorylase
kinase are passed through a Biogel column, an active fraction containing only the β-subunit can be isolated. This is the most convincing
proof that β represents the catalytic subunit; furthermore, this fraction is inhibited by EDTA, indicating that the β-subunit also requires
calcium.

No phosphorylation of dogfish phosphorylase kinase in either the α or
β-subunit could be demonstrated (Fig. 6). While rabbit muscle phosphorylase kinase is rapidly activated or inhibited by protein kinase or
phosphatase, no change in the activity of the dogfish enzyme is observed. In this instance, therefore, activation and inhibition seem to
rely solely on the availability of calcium ions (Pocinwong, 1975).

Since no phosphorylation of the α-chain occurs in the dogfish enzyme,
it would appear that this subunit first became incorporated into the
kinase complex simply as a regulatory component, and this might still
be its main function. Binding of Ca^{2+} to the β-subunit must reduce the
interaction between the catalytic and regulatory components, thereby
allowing enzymatic activity to express itself. In the mammalian system,
phosphorylation of the β-subunit further relaxes the complex and pro-

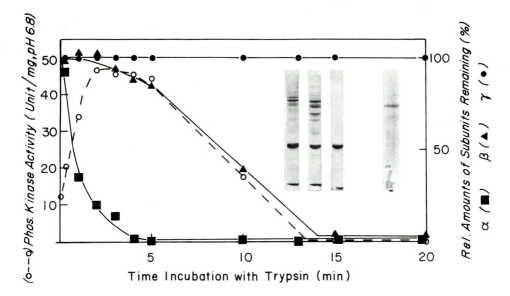

Fig. 5. Changes in activity of dogfish phosphorylase kinase by limited proteolysis. Trypsin (3.8 μg/ml) was added to phosphorylase kinase (1.9 mg/ml); at various times, samples were removed, treated with soybean trypsin inhibitor (50 μg/ml) and examined for kinase activity and subunit distribution. Three gels at various time intervals of trypsin treatment are shown: (left to right) 5 sec; 3 min; and 15 min. The gel on the far right is the isolated β-subunit

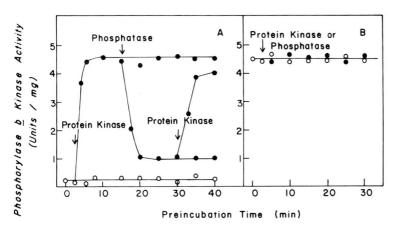

Fig. 6A and B. Effect of protein kinase and phosphatase on rabbit (A) and dogfish (B) phosphorylase kinase at pH 7.0, 20°, in the presence of 0.8 μM c-AMP, 2 mM Mg^{2+} and 0.6 mM ATP

vides for an additional increase in activity. At which point in time did the phosphorylation of the α-subunit come into play and what was the original purpose of this reaction? Was it to provide for further structural relaxation or flexibility in the modulation of enzyme activity, or to assist in the dephosphorylation of the β-subunit along the lines proposed in the "second site of phosphorylation" hypothesis?

We always wondered why the enzyme needed to incorporate into its structure a component as large as the α-subunit; surely not to provide,
originally, for a "second site" of phosphorylation, because there is
no way it could know that, several hundred millions year hence, it
would have to fulfill this function. Enzymes are smart, indeed, but not
that smart. Unless, of course, phosphorylation of the α and β-subunits
also occurred at that time, but was lost in the course of dogfish evolution.

The conversion of phosphorylase b to a catalyzed by phosphorylase
kinase in the presence of Ca, Mg and ATP is instantaneously blocked
by EGTA or purified preparations of fragmented sarcoplasmic reticulum
but resumes upon re-addition of an excess of Ca^{2+} ions. The dissociation constant for calcium (5×10^{-7} M) is of the same order of magnitude as that for the high affinity binding sites of TN-C (Potter and
Gergely, 1973).

II. The γ-Subunit

What about the γ-subunits? The enzyme was known to aggregate under
certain conditions, but the first indication of what might be taking
place came from electron microscopy of the dogfish enzyme performed
by Breck Byers: long filaments rather similar to those expected from
F-actin were observed (Fig. 7). And, of course, both the γ-subunit and

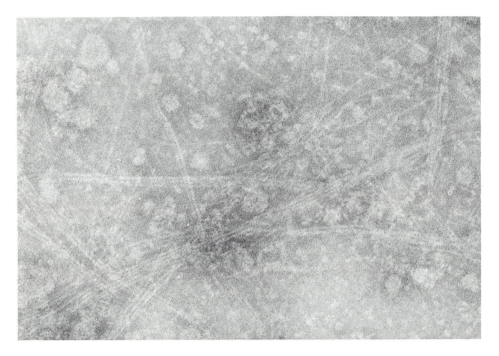

Fig. 7. Electron micrograph of dogfish phosphorylase kinase negatively
stained by uranyl acetate. (Performed by Dr. Breck Byers)

G-actin have a MW of ca. 45,000. Furthermore, purified γ-subunit migrated as a single band on SDS gel electrophoresis when mixed to purified dogfish actin. Finally, amino acid analysis of both proteins gave

Table 3. Amino acid analysis of dogfish actin and dogfish γ-subunit

Residue	Amino acid residue / 45,000 g	
	Dogfish γ	Dogfish actin
Lys	22	22
His	9	8
Arg	19	18
Asp	37	38
Thr	27	27
Ser	27	27
Glu	45	46
Pro	19	19
Gly	31	32
Ala	31	33
Val	23	22
Met	12	12
Ile	30	30
Leu	32	32
Tyr	17	17
Phe	14	13
1/2 Cys	5	5
Trp	4	4
3-Me His	0	1

identical compositions within the limits of error of the procedure (Table 3) with only one distinct difference: whereas dogfish actin, like most actin molecules so far investigated, contains one residue of 3-methylhistidine, the γ-subunit contains none.

The similarity between the γ-subunit of phosphorylase kinase and actin is not as apparent in the rabbit: whereas actin seems to have changed very little in the several hundred million years separating the two species, rather large differences are observed between the amino acid compositions of the dogfish and rabbit γ-subunits; furthermore, the latter is distinctly smaller (MW ca. 41,000).

What is the meaning of this surprising observation? Of course, many further experiments will have to be performed to confirm these data and establish their significance. Most of the structural proteins of dogfish muscle (troponin and its subunits, tropomyosin, actin, myosin and its light-chains etc.) have been purified and their interaction with phosphorylase kinase must be investigated. Does the γ-subunit affect myosin ATPase; is its aggregation subjected to the same effectors that determine the G to F-actin interconversion? Can the γ-subunit be methylated by S-adenosyl-methionine and if so, would this affect its chemical, physical and enzymatic properties?

No matter what, there must be some basic reason why the γ-subunit should be so closely related to actin. Perhaps its role is one of recognition;

148

that is, it could allow the enzyme to position itself properly within the myofibrils, for instance along the thin filaments. From recent work, we know that TN-I and TN-T can be phosphorylated by various protein kinases (Stull et al., 1972; Perry and Cole, 1974). Can the γ-subunit serve to position the enzyme next to the troponin molecules? Can the enzyme exchange its γ-subunits for those of G-actin to allow for this interaction? But here again, we are running into conceptual difficulties: if this were the sole function of the γ-subunit, i.e. to allow for a phosphorylation of troponin - how is it that no such phosphorylation was ever observed in dogfish muscle? Antibodies are being prepared against intact phosphorylase kinase, its γ-subunit and actin for intracellular localization; will these affect the subcellular structure of skeletal muscle and its contractility?

In summary, it can be said that in the white skeletal muscle of a primitive vertebrate such as the dogfish, there seems to be no direct hormonal intervention in the control of carbohydrate metabolism. There is no phosphorylation of protein kinase or phosphorylase kinase and, as far as could be determined, no production of c-AMP in response to epinephrine. The only protein kinase activity identified was independent of c-AMP. Furthermore, no phosphorylation of structural proteins such as TN-I, TN-T or the DTNB light-chain of myosin was observed in the presence of dogfish kinase (Malencik et al., 1975). On the other hand, these structural proteins were phosphorylated by various mammalian protein kinases as schematized in Fig. 8

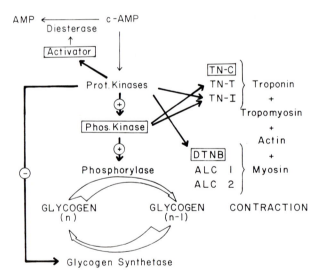

Fig. 8. Schematic representation of the phosphorylation of the regulatory enzymes of glycogen metabolism and muscle structural proteins. The bold arrows represent phosphorylation; the symbols (+ or -) represent activation or inactivation respectively, and the Ca^{2+}-binding proteins are framed

Control of carbohydrate metabolism in dogfish muscle must therefore rely essentially on the release of calcium from the sarcoplasmic reticulum following nerve impulse (Fig. 9). As the concentration of Ca^{2+} increases in the sarcoplasm, the metal ion will bind to troponin and phosphorylase kinase and trigger simultaneously contraction and glycogen breakdown.

For the concerted initiation of glycolysis and contraction, Ca^{2+} must bind simultaneously to TN-C and phosphorylase kinase as its concentration increases in the myoplasm. In the rabbit this requirement is clearly satisfied (Fig. 10A) as shown by work carried out in collaboration with a group in the Department of Physiology, University of Washington

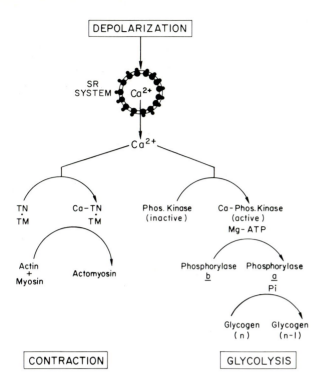

Fig. 9. Concerted regu-
lation of glycolysis
and muscle contraction
through Ca^{2+}

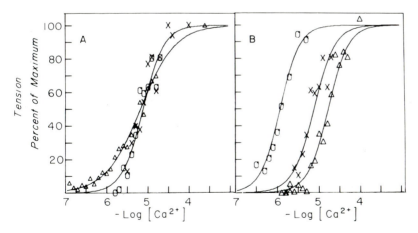

Fig. 10. (A) Tension (□) generated in rabbit skinned white muscle fibers
(*Adductor magnus*) as a function of Ca^{2+} concentration at 20°, pH 7.0.
Also plotted are (×) the binding of Ca^{2+} to TN-C and (△) the activation
of phosphorylase kinase. Solutions contain 70 mM K^+, 1 mM Mg^{2+}, 2 mM
$Mg-ATP^{-2}$, 15 mM phosphocreatine anion, 7 mM EGTA, 3 mg/ml of creatine
kinase and imidazole-propionate to maintain ionic strength (Kerrick
and Donaldson, 1972). (B) Tension curves for white skinned skeletal
muscle fibers of the dogfish (□), rabbit (×), and chicken (△) as a
function of Ca^{2+} concentration. Conditions as in Fig. A

(Kerrick and Krasner, 1975; Kerrick and Donaldson, 1972) using skinned white skeletal muscle fibers which lack a functionally intact sarcolemma. In such preparations, the ionic environment surrounding the contractile proteins can be perfectly controlled while isometric tension is monitored. Tension developed by the skinned muscle fibers as a function of Ca^{2+} concentration follows the uptake of this metal ion by both TN-C and phosphorylase kinase. In the latter case, computer fitting of the data indicates a single class of Ca^{2+} binding sites with a K_D of ca. 6 μM.

In addition to evolutionary differences in the control of glycogen breakdown, there are also variations in the relationship between Ca^{2+} concentration and tension for various species. As can be seen from Fig. 10B, dogfish muscle is 5-6 times more sensitive to Ca^{2+} than rabbit muscle, itself more sensitive than chicken muscle. It will be interesting to determine whether, in both cases, binding of Ca^{2+} to TN-C and phosphorylase kinase follows the same pattern.

E. The Low Molecular Weight Ca-binding Proteins

There was another reason for looking at the dogfish: this organism, as well as other fish and amphibians, was known to contain large amounts of low MW Ca-binding proteins called parvalbumins (Pechère et al., 1973) to which no real physiological function has been attributed. In fact, the possibility that parvalbumins might have an important physiological function was questioned, since these proteins were thought to be absent from higher vertebrates.

When it was found that dogfish muscle protein kinase did not act on phosphorylase kinase, a search was undertaken for its natural substrate. A low MW phosphate-acceptor protein (Blum et al., 1974) was eventually identified which, at first, appeared to be related to the parvalbumins. However, when the same purification procedure was applied to rabbit muscle extracts, a pure classical parvalbumin was isolated. Incidentally, no phosphorylation of parvalbumins was demonstrated under any experimental conditions.

Purified parvalbumins were isolated from turtle, chicken and rabbit muscle (Lehky et al., 1974; Pechère, 1974). They all have very similar properties, with MW of ca. 12,000; they are water soluble; they display the same characteristic UV spectrum with little or no absorbance at 280 nm (due to the absence of tryptophan and the presence, at most, of one tyrosyl residue), and vibronic structures in the 260 nm region due to a high phenylalanine content; they all bind 2 g-atom of calcium/mol with a dissociation constant below 1 μM. Fig. 11 shows the total sequence of rabbit muscle parvalbumin determined in collaboration with the group of Hans Neurath (Enfield et al., 1975). All the residues involved in the binding of calcium (6 in the first Ca-binding site and 5 in the second) have been strictly conserved, as well as Arg-75 and Glu-81 responsible for the salt linkage needed to maintain the protein in the proper conformation (Kretsinger, 1972; Moews and Kretsinger, 1975). The extent of homology with fish parvalbumins is of the order of 80% and homology was known to extend to several muscle structural proteins such as TN-C and the light chains of myosin (Collins et al., 1973; Weeds and McLachlan, 1974). Parvalbumins are present predominantly in the white skeletal muscle, with little usually seen in red muscle; however, they cannot be considered only minor components. As shown in Table 4, there is approximately 600 mg of parvalbumin/kg in rabbit muscle (i.e. a molar ratio 2.5-times below that of TN-C), approximately

```
                                                   15
Ac-ALA-MET-THR-GLU-LEU-LEU-ASN-ALA-GLU-ASP-ILE-LYS-LYS-ALA-ILE-

16                                                 30
GLY-ALA-PHE-ALA-ALA-ALA-GLU-SER-PHE-ASP-HIS-LYS-LYS-PHE-PHE-

31                                                 45
GLN-MET-VAL-GLY-LEU-LYS-LYS-LYS-SER-THR-GLU-ASP-VAL-LYS-LYS-

46                                                 60
VAL-PHE-HIS-ILE-LEU-ASP-LYS-ASP-LYS-SER-GLY-PHE-ILE-GLU-GLU-

61                                                 75
GLU-GLU-LEU-GLY-PHE-ILE-LEU-LYS-GLY-PHE-SER-PRO-ASP-ALA-ARG-

76                                                 90
ASP-LEU-SER-VAL-LYS-GLU-THR-LYS-THR-LEU-MET-ALA-ALA-GLY-ASP-

91                                                 105
LYS-ASP-GLY-ASP-GLY-LYS-ILE-GLY-ALA-ASP-GLU-PHE-SER-THR-LEU-

106         109
VAL-SER-GLU-SER
```

Fig. 11. Amino acid sequence of rabbit parvalbumin (Enfield et al., 1975). The residues implicated in Ca^{2+} binding of Carp III parvalbumin are framed; those corresponding to the hydrophobic core are underlined (Moews and Kretsinger, 1975)

3 times less in the chicken but 10 times more in the turtle. In rabbit muscle, this corresponds to a potential binding of Ca^{2+} of 0.14 mM.

Whether or not the function of parvalbumin is related to the contractile process or to calcium transport, for instance, we do not know.

From experiments carried out on skinned red and white muscle fibers, we know that Ca^{2+}-containing parvalbumin at concentrations as high as 2.5 mM does not affect the Ca^{2+} dependence of tension development. On the other hand, the metal-free protein (just as EGTA) will block within seconds the tension transients induced by Ca^{2+} release from the sarcoplasmic reticulum following caffeine administration. But parvalbumins are also found in other organs such as the brain and the spinal cord; in this instance, could they be involved in synaptic transmission or axonal transport, or could they function in conjunction with the actin microfilaments ubiquitously present in animal cells?

While these studies have clarified certain aspects of the concerted regulation of metabolic processes, they seem to have raised more questions than they have answered. Hopefully, some of the problems outlined here will be solved in the not-too-distant future.

Acknowledgments. This work was supported by grants from the National Institutes of Health, PHS (AM 07902, 17081-01 and HL 135-17-04), and National Science Foundation (GB 3249), the Muscular Dystrophy Association of America, and the Fonds National Suisse de la Recherche Scientifique (3.725.72).

The authors would like to thank Dr. Breck Byers for the electron microscopy and the Department of Fisheries for the use of their facilities.

Table 4. Properties of some low MW calcium-binding proteins

Protein	Approximate MW	Approximate concentration[a] g/kg in tissue		Ca²⁺-binding		Approximate Ca^{2+} concentration mmol/l intracellular water[a]
				$\bar{n}$	Approximate K_D (μM)	
Troponin-C[b]	18,000	2.2	rabbit muscle	4	2 at ~ 0.1 2 at ~10	0.6
DTNB Myosin[c] Light chain	18,000	5.8	rabbit muscle	1	5	0.43
Parvalbumin[d,e]	12,000	0.6	rabbit muscle	2	0.5	0.14
Protein activator[f] of cyclic nucleotide phosphodiesterase	15,000	0.06	beef brain	2-4	1-10	0.02
Vitamin D dependent[g,h] calcium binding protein	12,000	0.1	beef intestine	2	0.5	0.02

a Assuming intracellular water at 75% of tissue wet weight.
b Potter and Gergely (1974).
c Morimoto and Harrington (1974).
d Pechère et al. (1973).
e Lehky et al. (1974).
f Lin et al. (1974).
g Fullmer and Wasserman (1973).
h Hitchman and Harrison (1972).

References

Adams, M.J., Ford, G.C., Koeboek, R., Lentz, P.J., Jr., McPherson, A., Jr., Rossman, M.G., Smiley, I.E., Scheritz, R.W., Wonacott, A.J.: Nature (Lond.) 227, 1o98 (1970).

Adler, S.P., Mangum, J.H., Magni, G., Stadtman, E.R.: In: Proc. 3rd Int. Converence on Metabolic Interconversion of Enzymes (eds. Fischer, E.H., Krebs, E.G., Neurath, H., Stadtman, E.R.), p. 221. Berlin-Heidelberg-New York: Springer 1973.

Blum, H.E., Pocinwong, S., Fischer, E.H.: Proc. Natl. Acad. Sci. U.S. 71, 2198 (1974).

Cohen, P.: Biochem. Soc. Symp. 39, 51 (1974).

Cohen, P., Antoniw, J.F., Davison, M., Taylor, C.: In: Proc. 3rd Int. Conference on Metabolic Interconversion of Enzymes (eds. Fischer, E.H., Krebs, E.G., Neurath, H., Stadtman, E.R.), p. 221. Berlin-Heidelberg-New York:Springer 1973.

Collins, J.H., Potter, J.D., Horn, M.J., Wilshire, G., Jackman, N.: FEBS Lett. 36, 268 (1973).

Enfield, D.L., Ericsson, L.H., Blum, H.E., Fischer, E.H., Neurath, H.: Proc. Natl. Acad. Sci. U.S., in press (1975).

Fischer, E.H., Heilmeyer, L.G., Haschke, R.H.: Curr. Top. Cell. Regul. 4, 211 (1971).

Fischer, E.H., Pocker, A., Saari, J.C.: Essays in Biochemistry 6, 23 (1970).

Fullmer, C.S., Wasserman, R.H.: Biochim. Biophys. Acta 317, 172 (1973).

Hayakawa, T., Perkins, J.P., Krebs, E.G.: Biochemistry 12, 574 (1973a).

Hayakawa, T., Perkins, J.P., Walsh, D.A., Krebs, E.G.: Biochemistry 12, 567 (1973b).

Heilmeyer, L.G., Meyer, F., Haschke, R.H., Fischer, E.H.: J. Biol. Chem. 245, 6649 (1970).

Hitchman, A.J.W., Harrison, J.E.: Can. J. Biochem. 50, 758 (1972).

Huttunen, J.K., Steinberg, D., Mayer, S.E.: Proc. Natl. Acad. Sci. U.S. 67, 290 (1970).

Kerrick, G.W., Donaldson, S.K.: Biochim. Biophys. Acta 275, 117 (1972).

Kerrick, G.W., Krasner, B.: J. Applied Physiol., in press (1975).

Krebs, E.G.: Curr. Top. Cell. Regul. 5, 99 (1972).

Kretsinger, R.H.: Nature New Biol. 240, 85 (1972).

Larner, J., Villar-Palasi, C.: Curr. Top. Cell. Regul. 3, 195 (1971).

Learch, K., Muir, L.W., Fischer, E.H.: Biochemistry, in press (1975).

Lehky, P., Blum, H.E., Stein, E.A., Fischer, E.H.: J. Biol. Chem. 249, 4332 (1974).

Lin, Y.M., Liu, Y.P., Cheung, W.Y.: J. Biol. Chem. 249, 4943 (1974).

Malencik, D.A., Heizmann, C.H., Fischer, E.H.: Biochemistry 14, 715 (1975).

Meyer, F., Heilmeyer, L.G., Haschke, R.H., Fischer, E.H.: J. Biol. Chem. 245, 6642 (1970).

Moews, P.C., Kretsinger, R.H.: J. Mol. Biol. 91, 201 (1975).

Morimoto, K., Harrington, W.F.: J. Mol. Biol. 88, 693 (1974).

Pechère, J.F.: C. R. Acad. Sci., Paris 278, 2577 (1974).

Pechère, J.F., Capony, J.P., Demaille, J.: Syst. Zool. 22, 533 (1973).

Perry, S.V., Cole, H.A.: Biochem. J. 141, 733 (1974).

Perutz, M.F., Kendrew, J.C., Watson, H.C., J. Mol. Biol. 13, 669 (1965).

Pocinwong, S.: Ph.D. Thesis, University of Washington (1975).

Potter, J.D., Gergely, J.: Fed. Proc. 33, 1465 (1973).

Stull, J.T., Brostrom, C.O., Krebs, E.G.: J. Biol. Chem. 247, 5272 (1972).

Titani, K., Cohen, P., Walsh, K.A., Neurath, H.N.: Submitted to FEBS Lett. (1975).

Walsh, K.A.: Methods Enzymol. 19, 41 (1970).

Weeds, A.G., McLachlan, A.D.: Nature New Biol. 252, 646 (1974).

Discussion

<u>Dr. Heilmeyer</u>: Phosphorylase kinase was thought to be a very specific enzyme using as substrate only phosphorylase <u>b</u>. Since it is now known to act also on other substrates, e.g. the troponin subunits, it was interesting to look at its intracellular localization. Therefore FITC-labelled antibodies against phosphorylase <u>b</u> and phosphorylase kinase were prepared and incubated with transverse sections of rabbit diaphragm. Fig. 1 shows the intracellular localization of phosphorylase

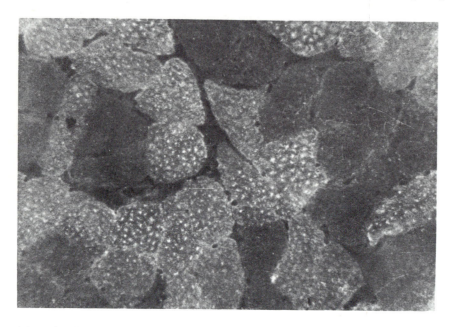

<u>Fig. 1</u>. Staining of rabbit diaphragm with anti-phosphorylase <u>b</u>. Antibodies against phosphorylase <u>b</u> were labelled with fluoresceineisothiocyanate (FITC) and incubated with thin sections of rabbit diaphragm. The fluorescence was observed in a Leitz Ortholux fluorescence microscope. For details see: W.H. Hörl, H.P. Jennissen, U. Gröschel Stewart, L.M.G. Heilmeyer, Jr.: Current Topics in Intracellular Regulation, Calcium Transport in Contraction and Secretion (eds. E. Carafoli, F. Clementi, A. Margreth). Amsterdam : ASP Biological and Medical Press B.V. North-Holland Division 1975

<u>b</u>; it appears to be concentrated in discrete areas, probably due to its presence in protein glycogen particles. In contrast, phosphorylase kinase is mainly located in the sarcolemma (Fig. 2) whereas the sarco-plasm is faintly stained. During centrifugation of isolated purified vesicles of the sarcoplasmic reticulum in a sucrose gradient as seen in Fig. 3, phosphorylase kinase, phosphorylase phosphatase and the Ca transport ATPase cosediment, whereas soluble enzymes such as phosphory-lase do not under these conditions. The influence of antibodies or phosphorylase <u>b</u> on the ATPase activity of the sarcoplasmic reticulum is shown in Fig. 4. No influence is observed on the rate of ATP split-ting in the absence and presence of Ca ions, whereas the further en-hancement of the ATPase activity by oxalate is completely suppressed This effect might indicate that phosphorylase kinase is involved in the regulation of the Ca pump in these membranes.

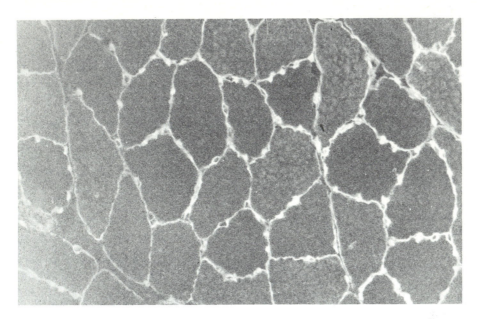

Fig. 2. Staining of rabbit diaphragm with anti-phosphorylase kinase. Labelling, staining and microscopy are described in the legend to Fig.1

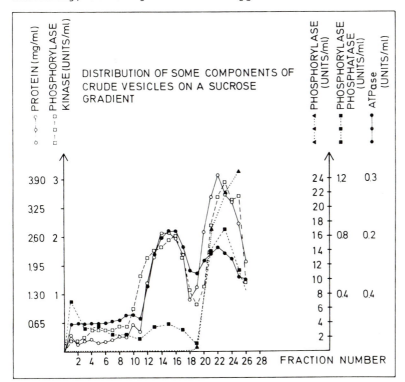

Fig. 3. Crude vesicles were layered on top of a linear sucrose gradient (0.3-0.9 M) and centrifuged at 8.000 × g for 60 min. Fractions were assayed for enzymatic activities

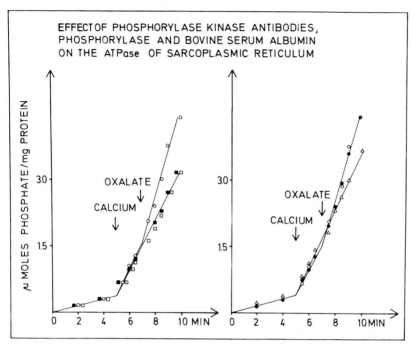

Fig. 4. Influence of anti-phosphorylase kinase on the Ca^{2+} transport ATPase; o——o represents control in presence of calcium and oxalate, ■——■ in absence of ocalate, □——□ in presence of 3 mg/ml anti-phosphorylase kinase. ●——● represents control in presence of 3 mg/ml bovine serum albumin; Δ——Δ in presence of 4.5 mg/ml phosphorylase b

Dr. Helmreich: Do you have some information on a possible interaction of dogfish phosphorylase kinase with myosin on HMM and does this affect either myosin ATPase or kinase activity? Or do you believe that phosphorylase kinase is attached to the cellular actin not related to muscle contraction?

Dr. Fischer: It appears that the γ-subunit of dogfish phosphorylase kinase interacts with HMM and myosin from dogfish, but not to the level obtained with purified actin. With respect to a recent report that actin inhibits DNase I, the γ-subunit is very much less effective, if at all. This inhibition might be correlated with the methylation of the protein which we propose to study later.

Dr. Wieland: Can the γ-subunit of rabbit phosphorylase kinase undergo polymerization to yield something similar to F-actin?

Dr. Fischer: We do not understand how this transition is brought about. For instance, it appears that more polymerization is observed with the intact enzyme than with the pure γ-subunit. Certain preparations of pure γ-subunit polymerize very poorly, while addition of HMM enhances polymerization. This might result from the fact that trypsin is used to isolate the γ-subunit. We have not yet investigated the conditions which affect polymerization and we would certainly like to look at compounds such as phalloidin that you have studied.

Dr. Wieland: Can you separate the γ-subunit without loss of activity?

Dr. Fischer: Yes, since we can even get an active β-subunit. Once you begin to trypsinize this molecule you can pull the subunits apart rather easily; the pure β-subunit is quite unstable while the γ-subunit is much more stable.

Dr. Wieland: Does the combination of β- and γ-subunits yield an enzymatically active complex?

Dr. Fischer: Perhaps so, since once you have wiped out the α-subunit by limited proteolysis, you are left with a β-γ complex. But you do not know whether this complex remains as such during activity measurements.

Dr. Pette: Can the α-, β-subunits be bound to F-actin so that the enzyme is bound to the I filament?

Dr. Fischer: This has not been tried as yet.

Dr. Pette: I have a question to ask Dr. Heilmeyer. You mentioned that only a small fraction of the total phosphorylase kinase activity is present in the SR fraction. However, in the fluorescent pictures you showed, it appeared to me that this should be the main fraction of the enzyme. In addition, there was a faint cross-striation which you suggested was due to the SR or the T-system. In a paper from our Laboratory (G. Dölken, E. Leisner, D. Pette: Histochem. 43, 113-121, 1975), we have recently localized glycolytic enzymes using fluorescent anti-enzymes. We find that many of the glycolytic enzymes are localized in the isotropic zone, i.e. in the I band from where they can be washed out, some of them with high ionic strength. I wonder whether the cross-striation in your case is not due to a staining of the isotropic zone which would suggest the presence of phosphorylase kinase within the interfilamentary space at the site of the actin filaments?

Dr. Heilmeyer: The interpretation of the band pattern in our pictures is based on the distribution of the enzyme during extraction of muscle tissue. Phosphorylase kinase activity was not detectable in extensively washed myofibrils whereas it could not be completely removed from SR preparations under the same conditions. It is also unlikely that the enzyme is adsorbed unspecifically to these membranes since the amount of sedimentable kinase activity was not enhanced upon addition of purified phosphorylase kinase to the vesicles. It is difficult to estimate the relative proportion of enzyme present in the membrane fraction versus the sarcoplasm for the following reason: if the amount of the phosphorylase kinase present in the SR fraction is calculated from its activity, a negligible portion of the total SR proteins should be represented by the kinase, which should never be detectable in SDS gels. However, the characteristic bands for the α- and β-subunits could be demonstrated. It might be therefore that the enzyme present in the membrane fraction has not the same specific activity as the soluble enzyme.

Dr. Jennissen: We have studied the relative turnover rates of the phosphorylase kinase subunits in rabbit and we have found that the γ-subunit turns over independently from α and β. This could reflect the evolutionary history of this kinase in rabbit muscle. However, it means that intracellularly, the α- and β-subunits must exchange with the γ-subunit. Have your attempted to exchange the γ-subunit of the rabbit enzyme with that of the dogfish?

Dr. Fischer: We would like very much to do this experiment. The difficulty is that we have not been able to pull these enzymes apart under nondenaturing conditions. But one should try this experiment with labelled proteins; for example, we want to isolate label actin and γ-subunit from both rabbit and dogfish muscle to carry out this kind of experiment. In turn, let me ask you a question: does the γ-subunit turn over slower than the two others?

Dr. Jennissen: Yes, the α- and β-subunits turn over at about twice the rate observed for the γ-subunit.

Dr. Drabikowski: What is the nucleotide content in the γ-subunit?

Dr. Fischer: Until now, we have not been able to show a strict stoichiometric binding of nucleotide to actin treated in the same way as the γ-subunit, that is, with trypsin and EDTA; the results with dogfish actin are of the order of 0.7 to 0.8 mol nucleotide per subunit, and only about 0.5 in the γ-subunit. But I would not trust these measurements at this time.

Dr. Weber: How much phosphorylase kinase is there in muscle?

Dr. Fischer: The concentration of phosphorylase kinase in muscle is in the order of 4 μM whereas that of troponin C is approximately 0.1 to 0.2 mM. The amount of Ca bound to phosphorylase kinase is very low in comparison with that bound to other muscle Ca-binding proteins. In rabbit muscle, the amount of Ca bound to parvalbumins is at least 1/4 that bound to TNC. But there is 10 times more parvalbumin in turtle muscle.

Dr. Demaille: In the case of rabbit skeletal muscle, the concentration of parvalbumin is 30-40 μmoles per kg muscle.

Dr. Gergely: I have a question to ask Dr. Heilmeyer. I was puzzled by the ATPase curves you showed. As Ca uptake occurs, the ATPase drops off when the vesicles get filled. By the addition of oxalate one expects another increase since due to the precipitation of Ca-oxalate the internal Ca concentration is lowered. In your experiment a steady increase was observed.

Dr. Heilmeyer: This is correct: in our experiments, the Ca uptake would cease only after about 10 min. We chose our conditions so that we were still working in the linear portion of the curve, i.e. within these 10 min.

Microtubules and Microfilaments

Microtubules: Inhibition of Spontaneous *in vitro* Assembly by Non-neural Cell Extracts

J. Bryan and B. W. Nagle

A. Introduction

Microtubules are a class of subcellular organelles found in all eukary-
otic organisms. In general these structures are long cylinders with a
diameter of approximately 24 nm and an apparently hollow core some
15 nm in diameter. Negative staining of microtubules with tannic acid
demonstrates that the 5 nm-thick tubule wall is composed of 13 protofilaments which are nearly parallel to the tubule long axis (see Tilney et al., 1974, also Nagano and Suzuki, 1975, and Hinkley and Burton,
1974). These protofilaments are composed of globular protein subunits
called tubulins (Mohri, 1968). Most microtubules appear to be composed
of two classes of tubulins, called α and β-tubulins, which normally
associate as a heterodimer (αβ) (Bryan and Wilson, 1971; Luduena et
al., 1974). The problems associated with packing tubulins into a microtubule have not yet been completely resolved; however, several quite
defined model structures have been proposed (see Amos and Klug, 1974;
Cohen et al., 1971; H. Erickson, 1974a; R.O. Erickson, 1973; Warner
and Meza, 1974; Warner and Satir, 1973). Further information on microtubule structure, including data on several polymorphic forms, can be
found in The Biology of Cytoplasmic Microtubules (published in the
Annals of the New York Academy of Sciences, 253, 1975).

In vivo, microtubules appear to be associated with a wide variety of
cellular processes including intracellular transport of material, chro-
mosome movements during mitosis, development and maintenance of cellular
form, cellular mobility and the possible control of the distribution of
certain cell surface receptors (see Bardele, 1973; Dustin, 1972; Bryan,
1974 for references). Each of these cellular processes is effected by
a class of drugs referred to as antitubulins. These drugs, which include colchicine, podophyllotoxin, vinblastine and their respective
derivatives share a common mode of action through their ability to dis-
rupt microtubules. Each of these agents is known to bind to the tubulin
dimer with high affinity and a reasonable amount of evidence is available on the stoichiometries and binding constants of each drug (reviewed
by Wilson and Bryan, 1975). A precise mechanism(s) of action for these
drugs is not yet available.

In the past two years, since the report of *in vitro* microtubule assembly by Weisenberg (1972), rapid progress has been made in understanding
the conditions required for the *in vitro* polymerization of brain tubulin.
Since this area is advancing quite rapidly, a review seems unwarranted;
rather, the interested reader should consult either The Biology of Cytoplasmic Microtubules or the ICN/UCLA Symposium on Microtubules (published in the Journal of Supramolecular Biology, 2, 1974).

The primary reason for using brain as a source of tubulin is simply
that spontaneous microtubule assembly will occur in suitably buffered
brain cytosol extracts warmed to 37°C, while such assembly has not been
observed under similar conditions in other types of cells, including

162

cultured cells or sea urchin eggs (see Bryan, 1975). In this paper we
would like to develop evidence that one reason for the failure to observe spontaneous microtubule assembly in non-neural tissue is the
presence of an inhibitor of microtubule assembly. We have chosen to
use marine eggs, because the available evidence indicates that they
contain substantial amounts of tubulin (Burnside et al., 1974; Raff
et al., 1973) and that cytosol preparations will form microtubules
if appropriate "seeds" are added (Burns and Starling, 1974; Weisenberg
and Rosenfeld, 1974). Our approach has been to use the assembly reaction of partially purified beef brain tubulin as an assay system to
monitor for potential inhibitors of tubulin assembly in non-neural extracts. Several types of cells have been looked at; for the sake of
brevity only data from sea urchin eggs will be presented. These eggs
contain a heat-stable, non-dialysable "factor" which effectively inhibits the spontaneous assembly of brain tubulin into microtubules by
blocking steps in the nucleation pathway. Using partially purified inhibitory factor, it is possible to show that tubulin solutions in which
spontaneous assembly is suppressed will assemble microtubules, presumably by elongation, if microtubule fragments are added. The initial
overall rate of this "nucleation" assembly is a linear function of the
number of short microtubule fragments added. This result has been used
to calculate an approximate average rate of elongation for an individual microtubule.

B. Methods

I. Tubulin Preparation

Bovine brain tubulin was prepared using minor modifications of the procedures described by Shelanski et al. (1973). Routinely, 1,000 gm
(wet weight) of fresh brain with the meninges dissected was washed free
of blood in an ice-cold saline solution, minced and suspended in 1,000
ml of ice-cold MES-Mg (0.1 M MES, 0.5 mM $MgCl_2$, pH adjusted to 6.5 at
23°C), then homogenized in a Waring blender for 2 minutes at half
speed. The homogenate was initially clarified by centrifugation at
15,000 × g for 15 min (4°C) and the supernatant recentrifuged at 50,000
× g for 75 min at 4°C. The final supernatant was diluted 1:1 with MES-
Mg containing 8 M glycerol and 2 mM GTP. (At this stage, the preparation can be stored at -20°C for up to a week without loss of ability
to assemble.) The diluted supernatant was then warmed to 37°C for 60
min and the polymerized tubules collected by centrifugation (50,000 × g
for 75 min at 25°C). The resulting pellet was resuspended in one-tenth
volume of prewarmed (25°C) MES-Mg buffer (without GTP or glycerol) by
gentle homogenization in a motor-driven teflon glass homogenizer. The
resuspended tubules were then cold-depolymerized at 0°C for 30 min,
then centrifuged at 100,000 × g for 60 min. The volume of the final
tubulin containg supernatants is measured and glycerol added (3.3 ml
per 10 ml of supernatant) with gentle stirring. These preparations are
stored at -20°C and used within 2-3 weeks. Following Borisy et al.
(1974), these are referred to as C_1S preparations. The purity of these
preparations has been examined using SDS gel electrophoresis. Approximately 90% of the stainable protein is α and β-tubulin or the HMW
components described by others (Borisy et al., 1974; Dentler et al.,
1975; Burns and Pollard, 1974; Gaskin et al., 1974b).

II. Assembly Assays

Three assay procedures have been used to monitor assembly. Where possible, the light-scattering assay developed by Gaskin et al. (1974a) has been employed. The results of this assay have been checked and normalized against centrifugation experiments where tubulin is assembled, the microtubules pelleted and protein concentrations determined by the Lowry procedure (Lowry et al., 1951). In some cases, when the scattering contributions of the fractions to be assayed are large, assembly has been monitored using viscosity measurements (Olmsted and Borisy, 1973).

Two types of tubulin preparations have been employed in the assembly assay; either the C_1S with 4 M glycerol present or a C_2S with or without glycerol present. The C_2S was obtained by adding GTP (to 1.0 mM) to a C_1S, then warming to $37^{\circ}C$ for 45 min, collecting the formed microtubules by centrifugation at 35,000 × g for 30 min ($25^{\circ}C$), resuspending the pellet in MES-Mg ($25^{\circ}C$), cooling at $0^{\circ}C$ for 30 min and finally clarifying by centrifugation at 35,000 × g for 30 min ($4^{\circ}C$). The results for both types of preparations are always qualitatively similar. In general, the initial rates of assembly in glycerol are slower at equivalent protein concentrations and temperatures and the time required to reach a stable plateau are greatly extended. Both preparations appear equally sensitive to dilution, addition of Ca^{++} or NaCl. The presence of glycerol, however, markedly decreased the cold lability and drug sensitivity of formed microtubules.

III. Preparation of Cell Extracts

Strongylocentrotus purpuratus eggs were collected, washed 3 times in large volumes of 19:1 (19 parts 0.53 M NaCl, 1 part 0.53 M KCl), then pelleted at approximately 800-1000 g. The pellet was resuspended in 5 volumes of ice-cold MES-Mg and gently homogenized in a teflon glass homogenizer. A high-speed supernatant was prepared by centrifugation of the homogenate at 100,000 × g for 60 min ($4^{\circ}C$). These cytosol preparations were tested for inhibitory activity or in later experiments were heated in a boiling-water bath for 15 min and recentrifuged at 100,000 × g for 60 min. Occasionally it was necessary to filter the first supernatants to remove lipids. In order to minimize the effects of residual salts, the high speed supernatants or boiled extracts were dialyzed (at $4^{\circ}C$) against several changes of MES-Mg buffer with or without added EGTA (1 mM), then clarified by centrifugation at 100,000 × g for 20 min when necessary.

C. Results

I. Assay System

Several features of the light-scattering assay and of spontaneous tubulin assembly are illustrated in the first two figures. The results are similar to those of Gaskin et al. (1974a). The time course of assembly for varying protein concentrations is shown in Fig. 1 for a C_2S preparation. After transfer to $30^{\circ}C$, there is an inital lag, followed by a pronounced increase in optical density (at 350 nm) until a stable plateau is reached. Fig. 2 illustrates the extent of assembly of microtubules at increasing protein concentrations determined with the centrifugation assay (see the legend of Fig. 2 for details). A comparison of the extent of assembly assayed by the centrifugation

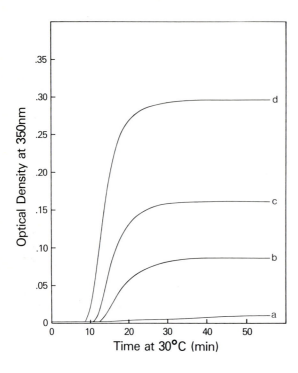

Fig. 1. Time course of microtubule assembly monitored using the light-scattering assay at 30°C and the C_2S preparations described in the Methods. The protein concentrations (mg/ml) are: a, 0.4; b, 1.0; c, 1.6; d, 2.7. The buffer is MES-Mg with a final GTP concentration of 0.5 mM

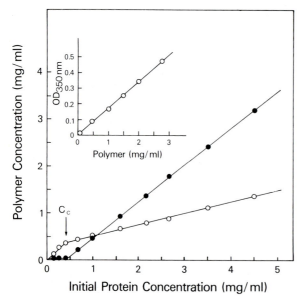

Fig. 2. Dependence of polymer formation on the initial concentration of protein. Two types of data are shown. In the main figure, the results of a centrifugation experiment are given. A C_2S preparation in MES-Mg with 0.5 mM GTP was diluted to various concentrations and warmed at 30°C for 60 min, then tubules were collected by centrifugation at 30,000 × g for 30 min. Protein concentrations were determined before and after centrifugation and used to calculate the protein concentration in microtubules (closed circles) and in soluble protein (open circles) at each initial protein concentration. The critical concentration (C_c) is indicated by the arrow. The inset is a comparison of the type of results given in Fig. 1 and Fig. 2, where the optical density at 350 nm after 60 min at 30°C is plotted against the concentration of polymer (mg/ml) obtained using the centrifugation procedure

procedure and by the light-scattering assay, shown in the inset in
Fig. 2, indicates that the O.D. increase is directly proportional to
the amount of polymer formed (see also Gaskin et al., 1974a).

The assembly reaction is completely reversible, if microtubules are
initially assembled at a high protein concentration, then used to prepare a concentration series by dilution of the formed microtubules
into pre-warmed buffer; the results are identical to those obtained
if tubules were allowed to assemble at a specified concentration. The
features of the assembly reaction are those predicted by Oosawa and
his colleagues (Oosawa and Kasai, 1962; Oosawa and Higashi, 1967) for
a condensation polymerization. Below a critical protein concentration
(C_c), little assembly can be detected; above the critical concentration,
polymer exists in equilibrium with unpolymerized tubulin. In these C_2S
preparations and to a greater degree in the C_1S preparations, there
is a constant increase in the concentration of unpolymerized protein
as the initial protein concentration is increased. Further purification by repeated assembly-disassembly cycles reduces the magnitude of
this increase to some extent (see Olmsted et al., 1974). The data presented, however, are typical for the tubulin preparations being used.
The C_cs are dependent upon several factors. The addition of inhibitors
of assembly such as the antitubulins, Ca^{++} or salt increases the apparent C_c in a systematic fashion which depends upon the concentration
of inhibitor present (Rodenberg, Eth and Bryan, unpublished). In addition, the C_c is related in a more complex fashion to the relative
amounts of tubulin in rings and dimers (Kirschner et al., 1974; H.
Erickson, 1974b; Olmsted et al., 1974). Preparations of tubulin which
contain a large percentage of tubulin in rings polymerize at significantly lower C_cs than preparations with lower ring contents. By intentionally enriching or depleting the relative concentration of rings,
the C_cs can be varied between 0.05 and >1.0 mg/ml. For the present experiments we have attempted to restrict the C_c range to between 0.4
and 0.6 mg/ml and will usually be comparing experiments done on the
same preparation.

II. Inhibitory Effect of Sea Urchin Extracts

Fig. 3 demonstrates the effect of increasing concentrations of the sea
urchin factor on the initial rates of assembly measured by the light-
scattering assay. In order to facilitate comparison, the data for the
cytosol and boiled extracts have been normalized against the percentage
(by volume) of sea urchin cytoplasm present in the reaction mixture.
This was done by measuring the initial egg pellet volume, then applying the appropriate corrections for subsequent dilution. Using this
normalization, it is clear that while there may be some loss of inhibitory activity on heating to 100°C, the bulk of the activity appears
to be heat-stable.

Fig. 4 illustrates the dependence of the inhibition on the initial concentration of tubulin. The data shown are the equilibrium polymer concentrations as a function of initial tubulin concentration for reactions done at various concentrations of boiled dialyzed extract. The
results were obtained using the light-scattering assay, then converting
to mg/ml of polymer using the calibration curve shown in the inset to
Fig. 2. The control, without added extract, is similar to the one shown
in Fig. 2, with a C_c = 0.55 mg/ml. The addition of the extract produces
a marked effect on the critical concentration and depresses the amount
of polymer found at all tubulin concentrations tested. Our initial analysis of this data indicates that at high tubulin concentration, the
observed decrease in polymer formed is approximately a linear function
of the concentration of boiled dialyzed extract added. The dependence

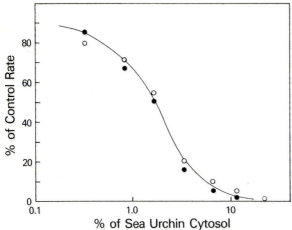

Fig. 3. Effect of added sea urchin extracts on assembly of beef brain tubulin. The results are plotted as percentages of the control rate of assembly versus the percentage (by volume) of sea urchin cytosol present in the reaction mixture. The rates used were the maximum rates observed for each sample. Open circles: added 100,000 × g cytosol preparation; closed circles: boiled extract. In the cytosol experiment, the 1% reaction mixture contained 225 µg of sea urchin protein while the boiled extract reaction mixture contained 45 µg of sea urchin protein. All reaction mixtures had tubulin (C_1S) at 1.2 mg/ml, GTP (1 mM) in MES-Mg buffer with 0.8 M glycerol. Temperature = $30^{o}C$

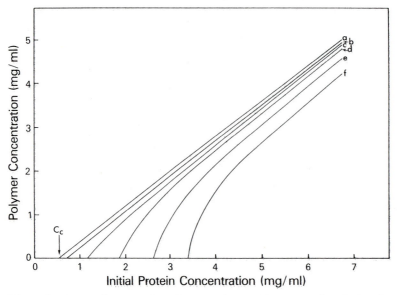

Fig. 4. Dependence of polymer formation on the inital protein concentration and on the concentration of boiled extract present in the reaction mixture (details see text). The tubulin used was a C_2S preparation in MES-Mg with 1.0 mM GTP and 1 mM EGTA. Curves: a, control; b, 10 µl; c, 20 µl; d, 50 µl; e, 100 µl; f, 200 µl of a boiled sea urchin extract (5.4 mg/ml) which had been dialyzed exhaustively against MES-Mg-EGTA. The actual points have been left out for the sake of clarity, but each line represents at least six measurements. Temperature = $30^{o}C$

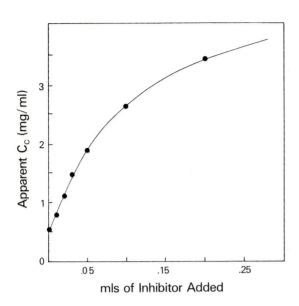

Fig. 5. Plot of the apparent
critical concentration (C_c)
as a function of the volume
of boiled extract added.
The conditions are those
given in Fig. 4

of the critical concentration on the concentration of extract is more
complex as shown in Fig. 5. The results indicate that there is an approach to saturation of the system with the inhibiting factor. This
series of experiments has been repeated using the centrifugation assay.
After correction for the soluble protein contributed by the added extract, the results are in reasonable agreement with those shown in
Fig. 4. In addition, qualitatively similar results were obtained for
preparations of C_1S (with 4 M glycerol present), C_2S (with or without
4 M glycerol present) and C_3S without glycerol.

III. Attempts to Relieve Inhibitory Activity

We have tried to relieve the inhibition by addition at zero time of
EGTA (1-10 mM), Mg^{++} (to 2.5 mM), excess GTP (to 5 mM), cyclic nucleotides, ATP, dithiothreitol, mercaptoethanol and 4 M glycerol. With the
exception of glycerol, none of these agents relieves the inhibitory
activity. Carrying out the entire assembly reaction in the presence of
4 M glycerol appears to reduce the sensitivity of assembly to the inhibitory factor by approximately a factor of 2 as determined by the
apparent critical concentrations. The inhibitory activity is still
present but higher concentrations (1.5-2-fold) of extract are required
to exert the same effect.

IV. Evidence that the Inhibitory Factor Blocks Nucleation

In order to test for the possibility that this factor was inhibiting
microtubule nucleation, the following experiments were done: (1) Extract was added at various times after assembly had started and preformed microtubules were present. (2) Fragments of sonicated microtubules, which could serve as potential nucleation sites, were added
to solutions of tubulin whose spontaneous assembly was suppressed by
the presence of the inhibitory factor.

The first class of experiments is complicated somewhat by the fact that
in order to add the extract, a dilution must be done which effects the

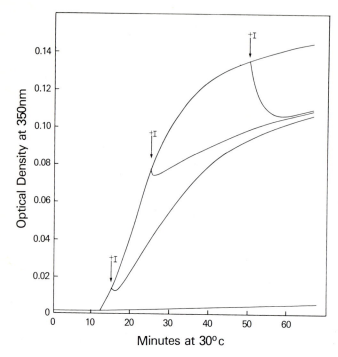

<u>Fig. 6.</u> The effects of addition of boiled dialyzed extract at different times during the assembly reaction. The reaction mixture contained 1.5 mg of C_1S protein in 1 ml of MES-Mg, 0.5 mM GTP and 4 M glycerol. At the times indicated, 0.1 ml of boiled extract was added and rapidly mixed with the sample. Base line: sample to which 0.1 ml of extract was added before transfer to 30ºC. The protein concentration of the boiled extract was 5.8 mg/ml in MES-Mg buffer

subsequent rate and extent of assembly. The results of one series of experiments are shown in Fig. 6. Addition of extract at zero time effectively suppresses spontaneous assembly throughout the time course of the experiment. Addition of the inhibitory factor at intermediate stages results in a decrease in rate approximately equal to that observed if an equal volume of buffer is added. This experiment has been repeated using progressively shorter times, but a minimum necessary preincubation time has not been firmly established. This particular set of experiments was done in the presence of 4 M glycerol containing buffers. Within 60 min, there is no clear-cut plateau but only a break in the original rapid assembly rate. In addition, the controls using freshly depolymerized tubulin with and without glycerol present have been done with qualitatively similar results. In all cases, if initiation or nucleation of microtubule growth is allowed to start, subsequent addition of the inhibitory factor is ineffective.

1. Addition of Sonicated Microtubule Fragments

The results of the second type of experiment are shown in Fig. 7. Using the type of information available from Figs. 4 and 5, a tubulin solution was prepared with sufficient boiled dialyzed extract present to insure that it was below its C_c. Aliquots of this solution were transferred to 30ºC and the $OD_{350 \, nm}$ monitored for 45 min or until the un-

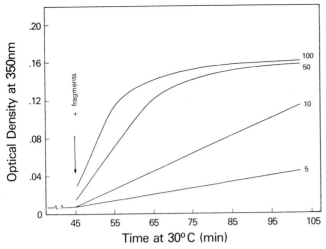

Fig. 7. The effect of adding sonicated microtubule fragments to supressed brain tubulin solutions. The reaction mixture contained tubulin (C_2S) at 2.7 mg/ml, GTP at 1 mM, 0.2 ml of boiled dialyzed extract (3 mg/ml of protein) in a total volume of 1.0 ml of MES-Mg buffer. The reaction mixtures were incubated at 30°C for 45 min before addition of the indicated volumes (in microliters) of sonicated fragments. See curve d, Fig. 1 for the behavior of a control solution under these conditions. The fragments were obtained by sonication of a control solution (2.7 mg/ml) for 15 sec

inhibited control had reached a plateau. At this time the control tubulin solution without added extract which had been allowed to assemble microtubules was sonicated for 15 sec (Branson Model W., at lowest setting) and aliquots were transferred into the inhibited solutions. The result is an immediate increase in optical density which starts, without lag, at the time of fragment addition and follows a simple expo-

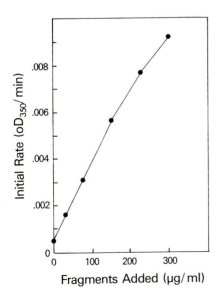

Fig. 8. Plot of the initial rate of polymer formation versus the final concentration of fragments added. The conditions for this experiment were essentially the same as those given in Fig. 7, using a somewhat higher protein concentration. The initial rates were determined from the linear portions of the assembly curves (see Fig. 7); the concentration of fragments added are a maximum estimate calculated assuming that all the protein in the addition was in fragments. The zero addition value was calculated from the small rise occuring over the initial 45 min incubation at 30°C. The maximum volume addition in this experiment was 0.1 ml

nential approach to the final equilibrium value. The observed initial rate of assembly is dependent on the final concentration of sonicated fragments added. These data are shown in Fig. 8 as a plot of the initial rates of assembly versus the final concentration of sonicated fragments added. The results indicate a linear dependence of the initial rate of net polymer formation on the number of added microtubule fragments. In these experiments the observed small non-linearity at higher concentration is probably due simply to the dilution of unpolymerized tubulin by addition of the fragment solutions, whose unpolymerized tubulin concentrations have been reduced by formation of microtubules.

In order to confirm that the assembly observed after fragment addition represents formation of microtubules, several types of experiments have been done. Direct E.M. observation of negatively-stained preparations indicated the presence of long microtubules at equilibrium. The addition of any of the antitubulins, Ca^{++} or NaCl effectively blocked the release of inhibition if added with the sonicated fragments. Finally, addition of Ca^{++}, NaCl or colchicine after equilibrium was reached produced a pronounced decrease in the amount of polymer.

V. Initial Fractionation Experiments

In order to begin to characterize the inhibitory activity and to rule out the possibility that the inhibition observed was simply due to the addition of "foreign" proteins, an initial fractionation of the inhibitor was done using gel filtration. A preparation of the boiled extract was chromatographed on a column of Agarose A-5 m (Bio-Rad Labora-

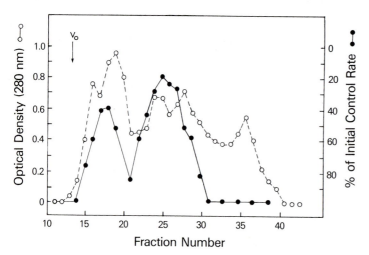

Fig. 9. Results of a gel filtration experiment using Agarose A-5 m (Bio-Rad Laboratories). 1 ml of a boiled dialyzed extract (5.2 mg of protein) was chromatographed on a 1 × 50 cm column of A-5 m equilibrated with MES-Mg. 1 ml fractions were collected and assayed for inhibitory activity by mixing 0.5 ml of each fraction with 0.5 ml of C_1S in MES-Mg with 4 M glycerol and 1 mM GTP (final tubulin concentration, 1.2 mg/ml), then monitoring the optical density at 350 nm after transfer to 30°C. Open circles: optical density at 280 nm of each fraction; closed circles: percentages of the initial control rate calculated using (experimental rate ÷ control rate) × 100. Using this normalization, 100% corresponds to no inhibition, while 0% is fully inhibited

tories) previously calibrated with a variety of globular protein molecular weight standards. The resulting fractions were tested for their ability to depress the initial rates of brain tubulin assembly and to lower the plateau values. The results using the rate data are shown in Fig. 9. Two peaks of inhibitory activity were found which do not correspond to the major protein distributions (as measured by $OD_{280 \ nm}$). The apparent molecular weights are 200,000 and 550,000. Since it is unlikely that the inhibitory activity is due to proteins, these numbers represent only relative size differences and not actual molecular weights.

D. Discussion

The evidence presented demonstrates that sea urchin egg cytosol preparations contain a factor (or factors) which can suppress the spontaneous assembly of brain tubulin. The chemical nature of this material is not yet clear, but a partial purification has been done by taking advantage of the heat stability. The inhibitory activity is non-dialysable, and can be separated into two fractions using agarose gel chromatography. It is not clear if these are distinct activities or simply different aggregation states. Results with other types of cells show that this activity is not unique to sea urchins, but can also be found in several cultured cell lines (Nagle and Bryan, in preparation). The precise mechanism of inhibition by the sea urchin factor is unclear. However, the experiments in Figs. 4 and 5 are consistent with the idea that the factor(s) binds either to a specific form of tubulin which determines the critical concentration or that the factor strips tubulin of some other molecule which determines the critical concentration of assembly. The available data (Fig. 5) suggest that it should be possible to saturate the system with the inhibitory factor and still observe polymerization at higher tubulin concentrations.

This argues that the inhibitory factor is not simply sequestering all of the available tubulin, but that even in the presence of excess inhibitory factor, there is a population of tubulin molecules which will form microtubules if their concentration is sufficiently high. Alternatively, if the inhibitory factor is removing a regulatory molecule, which lowers the critical concentration, then the saturation observed in Fig. 5 is an index of the intrinsic tubulin C_c.

Preliminary results using labelled colchicine to tag tubulin in the ring and dimer fractions suggest that colchicine binding to the ring tubulin is preferentially reduced or abolished by the presence of the inhibitory factor. This, and the fact that there is a marked dependence of the degree of inhibition of assembly (at constant extract concentration) on the relative concentration of rings suggests that the inhibitory factor alters ring function.

The idea that the inhibitory factor may alter ring function is made more attractive by the observations of Erickson (1974b) and Kirschner et al. (1974). These authors suggest that, *in vitro*, the rings probably unravel to form protofilaments which then aggregate laterally and eventually, in combination with dimers, form short closed tubules which serve as nucleation sites. It appears reasonable that the inhibitory factor may block one of these early steps requiring rings, since if the early steps are allowed to proceed or are by-passed with the addition of sonicated microtubule fragments, further assembly is possible. The inhibitory factor appears then to separate nucleation from microtubule growth or elongation.

While we would like to believe that this particular factor(s) is a biologically important regulator of tubulin *in vivo*, this view can be questioned. It is possible to argue the need for *in vivo* inhibitors of microtubule nucleation on the grounds that intracellular tubulin concentrations exceed the critical concentrations measured *in vitro*. Similarly, ultrastructural evidence such as that discussed by Tilney (1971) for the *in vivo* depolymerization and repolymerization of tubule-containing structures indicates little random spontaneous microtubule assembly. This evidence can, however, be interpreted in other ways. At the moment, there is no entirely convincing argument that this inhibitory activity functions *in situ*. From a practical point of view, however, these results will partially explain our own inability to obtain spontaneous, non-nucleated microtubule assembly in sea urchin eggs and cultured cell extracts and provide some basis for the observation of Burns and Starling (1974), that sea urchin extracts alone would not form polymers unless fragments of microtubules were added. In addition, the results suggest that either these inhibitory factors will have to be removed or nucleation sites will have to be added in order to study the control of microtubule assembly in cell extracts.

The inhibitor serves a second practical function. It can be used effectively to dissociate the nucleation process from microtubule elongation. This was shown here by the experiment (Figs. 7 and 8) designed to look at the relationship between the overall rate of assembly and the numbers of fragments added. The results show a linear dependence of the initial rate of assembly on the concentration of fragments added. This is in agreement with the results predicted by Oosawa and Higashi (1967) and allows a calculation of the rate of growth of an individual microtubule under the conditions used. The fragments added are several tenths of a micron long; for convenience we assume a length of 0.3 μ/fragment or approximately 500 tubulin dimers/fragment. We also assume that approximately 60% of the protein present in the sonicate is in microtubules initially and is in fragments after sonication. Using these numbers, a 1 μg addition of protein is equivalent to adding 6.5×10^9 fragments. From the slope of the curve in Fig. 8 and the calibration curve in Fig. 2 (inset), the overall rate of assembly is 0.17 μg per minute per μg of added fragments (0.17 μg/min/μg fragments) or approximately 150 dimers adding per microtubule per minute (150 dimers/microtubule/min). This represents a growth rate in terms of actual microtubule length of about 0.1 μ per min. This value is about 15 times lower than the figure determined for *in vivo* assembly (Goode, 1973). The numbers are not, however, comparable in a straightforward fashion since both temperature and tubulin concentration effect the process *in vitro*.

In summary, sea urchin egg cytosol preparations contain a heat-stable inhibitor of microtubule assembly which apparently functions by blocking some step in the nucleation pathway. It is not clear whether this inhibitory activity actually functions *in situ*, but *in vitro* it provides a useful method for suppressing spontaneous microtubule assembly. Using this inhibitory activity, it is possible to demonstrate that the initial rates of assembly in the absence of spontaneous nucleation are a linear function of the number of seeds or nuclei added and to make an estimate of the rate of microtubule growth under one set of conditions. These preparations are currently being used to try to evaluate the rate constants for elongation and to study the length distributions of microtubules under a variety of conditions.

Acknowledgment. This work was supported by NSF grants GB32287X and BMS74-20302.

References

Amos, L.A., Klug, A.: J. Cell Sci. 14, 523-549 (1974).
Bardele, C.F.: Cytobiologie 7, 442 (1973).
Borisy, G.G., Olmsted, J.B.: Science 177, 1196-1197 (1972).
Borisy, G.G., Olmsted, J.B., Marcum, J.M., Allen, C.: Fed. Proc. 33, 167-180 (1974).
Bryan, J.: Bioscience 24, 701 (1974).
Bryan, J.: American Zoologist 15, 649-660 (1975).
Bryan, J., Wilson, L.: Proc. Natl. Acad. Sci. 68, 1762-1766 (1971).
Burns, R.G., Pollard, T.D.: FEBS Letters 40, 274-280 (1974).
Burns, R.G., Starling, D.: J. Cell Sci. 14, 411-419 (1974).
Burnside, B., Kozak, C., Kafatos, F.C.: J. Cell Biol. 59, 755-762 (1973).
Cohen, C., Harrison, S., Stephens, R.E.: J. Mol. Biol. 59, 375 (1971).
Dentler, W.L., Granett, S., Rosenbaum, J.L.: J. Cell Biol. 65, 237-241 (1975).
Dustin, P., Jr.: Arch. Biol. (Liege) 83, 419 (1972).
Erickson, H.P.: J. Cell Biol. 60, 153-167 (1974a).
Erickson, H.P.: J. Supramol. Struct. 2, 393-411 (1974b).
Erickson, R.O.: Science 181, 705 (1973).
Gaskin, F., Cantor, C.R., Shelanski, M.L.: J. Mol. Biol. 89, 737-758 (1974a).
Gaskin, F., Kramer, S.B., Cantor, C.R., Adelstein, R., Shelanski, M.L.: FEBS Letters 40, 281-286 (1974b).
Goode, D.: J. Mol. Biol. 80, 531 (1973).
Hinkley, R.E., Burton, P.R.: J. Cell Biol. 63, 139a (1974).
Kirschner, M.W., Williams, R.C., Weingarten, M., Gerhart, J.C.: PNAS 71, 1159-1163 (1974).
Lowry, O.H., Rosebrough, N.J., Farr, A.L., Randall, R.J.: J. Biol. Chem. 193, 265 (1951).
Luduena, R., Wilson, L., Shooter, E.M.: J. Cell Biol. 63, 403a (1974).
Mohri, H.: Nature 217, 1053-1054 (1968).
Nagano, T., Suzuki, F.: J. Cell Biol. 64, 242-245 (1975).
Olmsted, J.B., Borisy, G.G.: Biochemistry 12, 4282-4289 (1973).
Olmsted, J.B., Marcum, J.M., Johnson, K.A., Allen, C., Borisy, G.G.: J. Supramol. Struct. 2, 429-450 (1974).
Oosawa, F., Higashi, S.: In: Progress in Theoretical Biology, Vol. I, pp. 79-164, 1967.
Oosawa, F., Kasai, M.: J. Mol. Biol. 4, 10-21 (1962).
Raff, R.A., Kaumeyer, J.F.: Develop. Biol. 32, 309-320 (1973).
Shelanski, M.L., Gaskin, F., Cantor, C.R.: PNAS 70, 765-768 (1973).
Tilney, L.G.: In: Origin and Continuity of Cell Organelles (eds. J. Reinert and H. Ursprung). Berlin-Heidelberg-New York: Springer 1971.
Tilney, L.G., Bryan, J., Bush, D.J., Fujiwara, K., Mooseker, M.S., Murphy, D.B., Snyder, D.H.: J. Cell Biol. 59, 267-275 (1973).
Warner, F.D., Meza, I.: J. Cell Sci. 15, 495-512 (1974).
Warner, F.D., Satir, P.: J. Cell Sci. 12, 313-326 (1973).
Weisenberg, R.C.: Science 177, 1104-1105 (1972).
Weisenberg, R.C., Rosenfeld, A.C.: J. Cell Biol. 64, 146-158 (1974).
Wilson, L., Bryan, J.: In: Advances in Cell and Molecular Biology, Vol. III, 21-72 (1975).

Discussion

Dr. Ponstingl: Several investigators find phospholipids associated with tubulin. Do these lipids play a role in microtubule assembly or are they just junk?

Dr. Bryan: A year ago we proposed that tubulin was in fact a lipoprotein, i.e. that it had lipids associated with it and that these may be important in polymerization. The rationale for doing this was that when we looked at crude extracts of chick embryo brain treated with lipases, assembly was completely blocked. This has been repeated a variety of times and it works quite well. The problem is that if one looks at tubulin that has been cycled a few times and treats it with lipases nothing happens. There is no effect whatsoever. If you cycle tubulin and look for lipids in the first cycle there are several moles of chloroform-methanol extractible phospholipids associated with the tubulin. If you look at it in the second cycle, the phospholipid content drops below stoichiometric amounts. In the third cycle it is still detectable but it is around 1/10 of a mol per tubulin subunit. It may be therefore just an impurity.

Dr. Ponstingl: The molar concentration of colchicine necessary to inhibit polymerization is one or two orders of magnitude lower than the concentration of tubulin. How does the substoichiometric inhibition fit into the current concept of assembly?

Dr. Bryan: What one is doing basically is poisoning a growth point and if you consider in that fashion one can understand a substoichiometric inhibition quite well. The problem is that it is very difficult to prove.

Dr. Ponstingl: I am not satisfied with that hypothesis, because you not only stop the growth, you depolymerize the tubules with colchicine.

Dr. Bryan: I do not know how great the emphasis is on that; we could perhaps pursue it in detail later. It is a question of the on and off rates of a dimer interaction with the growth point. One can devise conditions where one blocks the off-rate completely so that the tubules do not fall apart by growing them for example in high concentrations of glycerol or sucrose. When you do this and you add colchicine, what happens is that the growth stops; however, the tubules don't disassemble. I think it shows that if you can dissociate the on and off rates, it might solve this problem.

Dr. Eckstein: You left out deliberately the role of nucleotides. Can I ask whether the hydrolysis of GTP is essential for the reorganization of the proteins or whether the presence of GMP or GDP is accidental and has nothing to do with this process?

Dr. Bryan: At the moment it is a fairly dirty system. There are several papers which deal with this problem. What is turning out is that as one continues to purify it is becoming increasingly difficult to show that the nucleotide hydrolysis is in any may stoichiometrically related with putting a dimer into the tubule. It may be that it has only a regulatory function.

Microfilaments in Blood Platelets

E. F. Lüscher

A. Resting and Activated Blood Platelets

Blood platelets or thrombocytes are formed by fragmentation of the
cytoplasm of the megakaryocyte and therefore contain no nucleus. Nevertheless they are equipped with a wide variety of subcellular structures:
mitochondria, at least two types of storage organelles, and two internal
membrane systems, called "closed" and "open", the latter representing
invaginations of the plasma membrane which penetrate deeply into the
cytoplasm. A particularly prominent intracellular element is the microtubules, which circle the equatorial circumference of the cell in close
vicinity of the inner surface of the plasma membrane. It is generally
assumed that the microtubules act as a "cytoskeleton" and are responsible for the disc-shape of the circulating, "resting" platelet.

Platelets can be triggered into activity by an astonishing variety of
external agents, which for convenience may be subdivided into three
classes:

- Proteolytic enzymes, of which thrombin is most important. Other pro-
teases are trypsin, papain, and a variety of snake venom proteinases
(cf. Davey and Lüscher, 1967). It is generally assumed that the thrombin
substrate on the platelet surface is a glycoprotein without any relationship to fibrinogen (Phillips and Agin, 1974).

- "Large molecules", a quite heterogeneous class, including substances
such as collagen, aggregated immunoglobulins (IgG), and certain poly-
cations (see Lüscher et al., 1973).

- "Small", pharmacologically active molecules, such as adrenaline,
serotonin, vasopressin, and adenosine-5-diphosphate (ADP). ADP is a
particularly powerful inducer of platelet activation, depending for
activity on the presence of fibrinogen and, like other small molecules,
also on the presence of Ca^{2+}-ions. It is noteworthy that all these
materials are vasoactive, i.e. capable of stimulating the contraction
of smooth muscle.

The sequence of events which follows the exposure of platelets to an
activating agent may have some variation but in most cases is as follows:

- A rapid transformation of the disc-shaped platelet to a "spiny sphere".
This process, which takes place within a few seconds and without an increase in the cell volume, is linked to the appearance of thin "spikes"
which may be as long as several diameters of the original cell (Born,
1974). Rapid shape change involves the disappearance of the original
band of microtubules.

- A pronounced tendency towards aggregation, the extent of which depends
on the amount of inducing agent. With small amounts of inducer, aggre-

gation is spontaneously reversible; the platelets thereby assume again their original disc-shape. With more inducer, more dense aggregates will form and this is accompanied by the release reaction.

- The release reaction, consisting of the specific secretion of material which is contained in the storage organelles (cf. Holmsen et al., 1969). Human platelets release adrenaline, serotonin (5-hydroxytryptamine), adenine nucleotides (in particular ADP and ATP), K^+ and Ca^{2+}-ions, a complex of a heparin-neutralizing protein with chondroitin-sulphate A, as well as some other proteins, among them fibrinogen, and a glycoprotein. It is noteworthy that some of the released materials, such as ADP and the biogenic amines are themselves release inducers.

- The onset of contractile activity. Loose, voluminous platelet aggregates will spontaneously contract to a smaller volume and the long spikes formed by the platelets in the course of "rapid shape change" are completely withdrawn again. If activated platelets are incorporated in a fibrin clot, they will cause the spontaneous contraction of this clot, a phenomenon termed clot retraction.

All these last-mentioned manifestations of platelet activity quite clearly involve the presence in the cell of a contractile system.

B. The Platelet Contractile System

The presence in human platelets of an actomyosin-like contractile protein was first described in 1959 by Bettex-Galland and Lüscher and was confirmed shortly afterwards for pig platelets by Grette (1962). At that time, the existence of such proteins in non-muscular cells had already been established (cf. Hoffmann-Berling, 1961); however, none of them had been described in detail. Due to the fact that platelets are an easily obtainable starting material and contain large amounts of actomyosin (up to 15% of their total protein content), conditions for isolation and characterization appeared particularly favorable. Nonetheless, it is only in recent years that pure preparations have been obtained which allow an accurate assessment of the similarities and dissimilarities with the corresponding material from muscle.

It should be noted that platelet actomyosin was originally named thrombosthenin by its discoverers. In view of the fact that closely related, if not the same proteins are found in a variety of other non-muscular cells, the use of a special name for the platelet material is no longer justified.

I. The Components of Platelet Actomyosin

1. Platelet Actin (Thrombosthenin A)

Platelet actin is very similar to actin of muscular origin (cf. Bettex-Galland et al., 1962). It has a molecular weight of 43,000 and contains one methyl-histidine per mole (Adelstein and Conti, 1972; Booyse et al., 1973). As with muscle actin, the globular G-form will polymerize to the fibrillar F-form. With heavy meromyosin originating either from platelets or from striated muscle, the typical arrowhead configuration is formed (Behnke et al., 1971; Bettex-Galland et al., 1972).

Antisera prepared against actin (Lazarides and Weber, 1974) will cross-react not only with actins from non-muscle cells but with the actins from smooth and striated muscle as well. The same is true for the anti-actin autoantibody found in the sera of patients with chronic inflammatory hepatitis (Gabbiani et al., 1973). Also like muscle actin, platelet actin is a stimulator of the myosin ATPases of platelet or muscle origin (cf. Bettex-Galland and Lüscher, 1965).

It has been reported that platelet actin is present in two different forms, one of them being unable to polymerize to the F-form (Probst and Lüscher, 1972). Further evidence for two platelet actins has been provided by Abramovitz et al. (1972). At the present time it is difficult to draw any conclusions from these findings. It is interesting, though, that platelets contain much more actin than myosin.

Finally, it must be mentioned that platelet actin is cleaved by the enzyme thrombin, whereby an "actinopeptide" is released (Shainoff, 1973). Upon additon of 0.1 M KCl to thrombin-treated platelet G-actin, a coagulum forms. Again, the functional significance of these findings is not yet clear, since under *in vivo* conditions a direct contact between actin and thrombin, at least in the initial phase of platelet activation, is difficult to visualize.

2. Platelet Myosin (Thrombosthenin M)

Platelet myosin consists of two heavy chains (MW 200,000) and two pairs of light chains (MW 20,000 and 17,000 respectively) (Adelstein and Conti, 1974).

Booyse et al. (1971) have determined by ultracentrifugation a MW of 542,700. Smooth muscle myosin also has 2 light chains of the same molecular weight (Kendrick-Jones, 1973; Leger and Focant, 1973) which are both different from those of striated muscle. Therefore, the conclusion that platelet myosin is much more related to the corresponding material from smooth than from striated or cardiac muscle seems justified.

As a rule, myosin is isolated from platelets together with proteolytic split products corresponding to the "head" and "rod"-portions (Adelstein et al., 1971). According to Abramowitz et al. (1974) this represents a preparation artefact due to proteolysis in the course of prolonged storage of platelets.

There is not such wide immunological cross-reactivity with myosin as with actin. In fact, Adelstein (1975) reports that antibodies made against smooth muscle myosin will not cross-react with platelet myosin. The specificity of this antibody seems to be determined by the rod- and not by the head moiety of the molecule.

Platelet myosin is an ATPase which at high ionic strength in the presence of K^+ and EDTA, or of Ca^{2+}, has about the same activity as at low ionic strength in the presence of Mg^{2+}-ions and actin (Adelstein and Conti, 1974). The observed maximal values (1.4 µ moles P_i/mg protein/min) are about the same as those reported for smooth muscle myosin under comparable conditions.

Recently, the presence in platelets of an enzyme which specifically phosphorylates the 20,000 dalton light chain of myosin has been reported (Adelstein et al., 1973; Daniel et al., 1974). This platelet myosin light chain kinase is activated by Mg^{2+}. It can also phosphorylate the corresponding light chain in smooth muscle myosin but it is totally inactive with myosins from striated and cardiac muscle. There is no effect of phosphorylation on the myosin-ATPase activity measured at high salt concentration in the presence of either K^+ and EDTA, or Ca^{2+}-ions. However, when the phosphorylated myosin combines with actin, a 7- to 10-fold increase in the ATP-splitting activity is observed (cf. Adelstein, 1975). It certainly will be of great interest to see whether this phosphorylation with the concomitant increase in enzymatic activity gives platelet actomyosin altered properties with respect to actual contractile activity. In this context it is also of interest that Pollard (1975) claims that in order to have full ATPase activity in the actin-myosin system of platelets, cofactors are required. One might speculate that the phosphorylating system plays the role of such a cofactor.

Like platelet actin, platelet myosin is also a substrate for the enzyme thrombin (Cohen et al., 1969; Booyse et al., 1972). This proteolytic degradation does not interfere with the myosin-ATPase-activity. The functional significance of this process is still unclear.

II. Regulatory Proteins and "Relaxing Factor"

1. Platelet Tropomyosin was first described by Cohen and Cohen (1972) as a two-chain molecule with a subunit molecular weight of 30,000 daltons. Its paracrystals show a periodicity which is considerably shorter (345 Å) than the one observed with striated muscle tropomyosin (400 Å). This suggests that the platelet protein is a smaller molecule and in fact, tropomyosin from striated muscle has a slightly higher molecular weight (34,000 daltons).

2. Platelet Troponin. Platelets contain as yet ill-defined material which is capable of conferring Ca^{2+}-sensitivity to actomyosin from platelets and from rabbit muscle (Thorens et al., 1973). It seems likely enough that platelet actin and tropomyosin should be complemented with troponin, especially as there is conclusive evidence for the importance of Ca^{2+}-ions in the activation of platelet contractility. Recently Cohen et al. (1973) have provided additional evidence for the presence of troponin in human platelets.

3. "Relaxing Factor" was first described in platelets by Grette (1963). White and Krivit (1967) suggested that platelets might contain a membrane system comparable to the sarcoplasmic reticulum of muscle and later, Statland et al. (1969) demonstrated the uptake of calcium by membranous vesicles obtained from homogenized platelets. Later the identity of the calcium-accumulating vesicles with the so-called dense tubular system was clearly established (White, 1972).

III. Localization of the Contractile System of Platelets

In the resting platelet there is no evidence whatsoever for the presence, within the cytoplasmic space, of preformed filamentous structures which could be attributed to actomyosin. However, upon activation, thin filaments appear abundantly throughout the cytoplasm. Their dimensions as well as the fact that they combine with heavy meromyosin to typical arrowhead structures leave no doubt that they are F-actin (Shepro et al., 1969; Zucker-Franklin, 1969; Behnke et al., 1971; Zucker-Franklin and Grusky, 1972). Of particular interest is the observation that actin filaments are arranged in a rectangular pattern in close vicinity of the inner demarcation of the plasma membrane, including its invaginations (Zucker-Franklin, 1970). Since actin appears to be a tightly-bound membrane constituent (Podolsak, unpublished observation; Taylor et al., 1975) it is conceivable that this contractile network is intimately linked to morphological alterations of the cell surface.

Platelet myosin is much more difficult to discern on electron micrographs of activated platelets, although large amounts of actin filaments may be present. It is only with special techniques that myosin filaments, mostly in the form of aggregates, become visible (Bettex-Galland et al., 1969; Behnke et al., 1971; Rosenbluth, 1971; White, 1971; Zucker-Franklin and Grusky, 1972). According to Niederman and Pollard (1975) platelet myosin filaments have much smaller dimensions than those prepared from skeletal muscle. This may be part of the explanation for the poor expression of myosin in platelets.

Several authors have claimed that platelet actomyosin is also located on the outer side of the plasma membrane, where it would play an important role in aggregation and as a so-called "ecto-ATPase" (Salzman et al., 1966; Chambers et al., 1967; Booyse and Rafelson, 1969, 1972). On the other hand, it has been pointed out that a Ca^{2+}- and ATP-dependent system appears rather meaningless in a physiological suspension medium; provided ATP was available at all, it should remain in a constantly contracted state (Lüscher and Bettex-Galland, 1972; Lüscher et al., 1972). Immunological techniques, provided antisera are used which are not contaminated with antibodies directed against membrane components, show no evidence for actomyosin located outside the resting platelet. Thus, the anti-actin antiserum from patients with chronic inflammatory hepatitis does not combine with intact platelets, although it will react with damaged cells (Gabbiani et al., 1973). Pollard (1975), working with a highly-purified platelet myosin antigen, was unable to detect fixation of the corresponding antibody on the surface of intact platelets.

In summary, platelet actomyosin is localized in the resting platelet in the cytoplasm. It appears to be present in a dispersed form, perhaps consisting of oligomers of actin and myosin which escape electron microscopical observation. Upon activation, a polymerization process is initiated, leading to large amounts of F-actin filaments, again in the cytoplasmic matrix, or in the form of submembranous filaments adhering to the inner surface of the plasma membrane.

In the resting platelet, no actomyosin is found outside the plasma membrane. This does not exclude the possibility that contractile protein becomes accessible from the outside in the course of membrane rearrangements which accompany platelet activation.

C. Mode of Activation of the Contractile System

From the composition and properties of the platelet's contractile system it follows that activation must be brought about by the availability of Ca^{2+}-ions in the cytoplasm. In theory, this calcium can either be mobilized from internal storage organelles - the "relaxing factor" system - or, under physiological conditions, from the outside through a plasma membrane which acquires calcium permeability.

Rapid shape change, i.e. the formation of long spikes linked to the disc-sphere transformation and followed by the retraction of these pseudopodes, must be looked upon as a manifestation of contractile activity. Since this phenomenon is also observed, e.g. after the addition of ADP to platelets in an EDTA-containing medium, it is obvious that in this case the Ca^{2+}-ions must originate from internal sources.

It is of particular interest that the disc-sphere transformation involves the disappearance of the ring of microtubules, which is either dissolved or shifted to a more central position. Microtubules are destroyed by Ca^{2+}-ions (Weisenberg, 1972); their disappearance therefore is an indicator for the release of calcium into the cytoplasm.

It should be noted that this inverse dependence on calcium of microtubules and contractile filaments is the only parameter which links these two systems. Platelet microtubules are not contractile elements and speculations as to the identity of their bulding unit - tubulin - with actin are no longer tenable (cf. Puszkin et al., 1969). In fact purified tubulin is not an activator of the myosin ATPase, as claimed by these last-mentioned authors (Käser-Glanzmann et al., 1975).

The platelet plasma membrane normally has a very low permeability for Ca^{2+}-ions. However, upon stimulation, e.g. by thrombin, a rapid influx of the cation is observed (cf. Lüscher and Massini, 1975). In fact, clot retraction, a typical manifestation of platelet contractility, can be initiated and stopped at will by variation of the external calcium concentration (Cohen and de Vries, 1973). This shows that Ca^{2+}-ions can move in both directions through the activated membrane.

Ionophores offer the possibility of transporting Ca^{2+}-ions passively through membranes. Accordingly, clot retraction can also be induced in a platelet-containing fibrin clot produced in the presence of Ca^{2+}-ions with Reptilase [+], a fibrinogen-clotting snake venom enzyme which by itself does not activate platelets (Massini and Lüscher, 1974).

It is of interest that ionophores will also induce the release reaction (Feinman and Detwiler, 1974; Massini and Lüscher, 1974), although there is ample evidence that this manifestation of platelet activity is independent of the contractile system (White, 1971).

Thus, it is clearly established that activation of the platelet is achieved either by the mobilization of internal, organelle-bound calcium or by the influx of Ca^{2+}-ions through the permeable plasma membrane. Since manifestations of contractility are also observed in a medium containing chelators of Ca^{2+}-ions, such as EGTA or EDTA, it must be postulated that a signal for the internal release of the cation originates from the plasma membrane as a consequence of its contact with a suitable inducer of activity. At present, nothing definite is known about the nature of this signal, although there is some likelihood that cyclic AMP or cyclic GMP might play the role of a "second messenger" (cf. Marquis et al., 1970; Salzman and Levine, 1971, but also Haslam, 1975).

D. The Physiological Significance of the Platelet Contractile System

It has been pointed out before that very probably the early morphological alterations observed in activated platelets are due to the contraction-relaxation, mainly of membrane-associated actomyosin. The formation of long spikes in the course of "rapid shape change" is a particularly intriguing phenomenon, the explanation of which is as yet uncertain. One possibility would be that a contraction of the submembranous filaments leads to the build-up of internal pressure, which is released by pushing out pseudopodes at sites of minor resistance in the membrane. An equally attractive hypothesis is based on the observation that actomyosin is very effective in initiating and sustaining cytoplasmic streaming. Thus the transport of liquid in glass capillaries containing platelet actomyosin has been observed (Cohen et al., 1974); it is tempting to speculate that the pseudopode acts as a "capillary", thereby causing its own prolongation. Quite generally speaking, the role of actomyosin in cytoplasmic streaming may be of great importance for the understanding of morphological cell alterations and perhaps also for cell motility. In this context it is of interest that Wu Loewenhaupt et al. (1973) claim that platelets can actively migrate in response to the chemotactic stimulus exerted by collagen.

The major physiological role of blood platelets consists in their participation in the haemostatic process. They will immediately adhere to the site of an endothelial injury, undergo activation (mainly by contact with collagen), in the course of which they become themselves foci for the attachment of new, circulating platelets. By continuing repetition of this process, a voluminous aggregate will form in a short time. Microscopical observation shows that these primary aggregates are quite fragile and unable to withstand the eroding forces of the outflowing blood. This changes suddenly: the platelet mass becomes solid and acts as an effective barrier against further blood loss. It should be noted that this process, which determines the spontaneous arrest of bleeding from smaller wounds, is completely independent of fibrin formation which sets in later on.

There can be little doubt that this essential "consolidation" of the so-called haemostatic plug is due to the contraction of the loose primary aggregate. In fact, the contraction of platelet aggregates can easily be observed *in vitro* (cf. Sokal, 1960). The same process most likely is also responsible for the solidification of intravascular platelet masses which, as a "white thrombus", can obstruct blood circulation, a phenomenon which is of primary importance in arterial thrombosis.

Lastly, the most easily observable manifestation of platelet contractile activity must be mentioned: clot retraction. Known and investigated over a long period, this process is still full of puzzles. The most plausible explanation of the spontaneous contraction of a platelet-containing, thrombin-induced fibrin clot (in the presence of calcium ions) is based on the contraction of the long pseudopodes which form during early shape change of the platelets, which attach themselves to fibrin fibers and carry them along when their own contraction is initiated. The coincidence of the onset of retraction with the appearance of microfilaments in platelets is well established (Chao et al., 1970).

The platelet may appear as a very special cell - it is unique with respect to its formation and plays a very specific role in haemostasis and blood coagulation. It is becoming more and more clear, though, that contractile systems such as the one found in platelets are widely distributed in many other cell types. Therefore observations made on the easily accessible platelet actomyosin may shed light on other comparable systems which, because of scarcity of material and for other reasons, are more difficult to study. There is in fact every reason to believe that the platelet is the perfect model of a contractile, non-muscular cell.

References

Abramowitz, J., Stracher, A., Detwiler, T.C.: Biochim. Biophys. Res. Comm. 49, 958-963 (1972).

182

Abramowitz, J., Stracher, A., Detwiler, T.C.: J. Clin. Invest. <u>53</u>, 1493-1496 (1974).
Adelstein, R.S.: Haemostasis 5, in press (1975).
Adelstein, R.S., Conti, M.A.: Cold Spring Harbor Symp. Quant. Biol. <u>37</u>, 599-606 (1972).
Adelstein, R.S., Conti, M.A., Anderson, W., Jr.: Proc. Nat. Acad. Sci. USA <u>70</u>, 3115-3119 (1973).
Adelstein, R.S., Pollard, T.D., Kuehl, W.M.: Proc. Nat. Acad. Sci. USA <u>68</u>, 2703-2707 (1971).
Behnke, O., Kristensen, B.I., Engdahl-Nielsen, L.: J. Ultrastruct. Res. <u>37</u>, 351-367 (1971).
Bettex-Galland, M., Lüscher, E.F.: Nature (Lond.) <u>184</u>, 276-277 (1959).
Bettex-Galland, M., Lüscher, E.F.: Adv. Protein Chem. <u>20</u>, 1-35 (1965).
Bettex-Galland, M., Lüscher, E.F., Weibel, E.R.: Thromb. Diath. Haemorrh. <u>22</u>, 431-449 (1969).
Bettex-Galland, M., Portzehl, H., Lüscher, E.F.: Nature (Lond.) <u>193</u>, 777-778 (1962).
Bettex-Galland, M., Probst, E., Behnke, O.: J. Mol. Biol. <u>68</u>, 533-535 (1972).
Booyse, F.M., Hoveke, T.P., Kisieleski, D., Rafelson, M.E., Jr.: Microvascular Res. <u>4</u>, 199-206 (1972).
Booyse, F.M., Hoveke, T.P., Rafelson, M.E., Jr.: J. Biol. Chem. <u>248</u>, 4083-4091 (1973).
Booyse, F.M., Hoveke, T.P., Zschokke, D., Rafelson, M.E., Jr.: J. Biol. Chem. <u>246</u>, 4291-4297 (1971).
Booyse, F.M., Rafelson, M.E., Jr.: Blood <u>33</u>, 100-103 (1969).
Booyse, F.M., Rafelson, M.E., Jr.: Microvascular Res. <u>4</u>, 207-213 (1972).
Born, G.V.R.: In: Erythrocytes, Thrombocytes, Leukocytes (eds. Gerlach, E., Moser, K., Deutsch, E., Wilmans, W.), pp. 253-257. Stuttgart: Thieme 1972.
Chambers, D.A., Salzman, E.W., Neri, L.L.: Arch. Biochem Biophys. <u>119</u>, 173-178 (1967).
Chao, F.C., Shepro, D., Yao, F.: Microvascular Res. <u>2</u>, 61-66 (1970).
Cohen, I., Bohak, Z., de Vries, A., Katchalski, E.: Europ. J. Biochem. <u>10</u>, 388-394 (1969).
Cohen, I., Cohen, C.: J. Molec. Biol. <u>68</u>, 383-390 (1972).
Cohen, I., Kaminski, E., de Vries, A.: FEBS Letters <u>34</u>, 315-318 (1973).
Cohen, I., Tirosh, R., Oplatka, A.: Pflügers Arch. ges. Physiol. <u>352</u>, 81-85 (1974).
Cohen, I., de Vries, A.: Nature (Lond.) <u>246</u>, 32-37 (1973).
Daniel, J., Conti, M.A., Adelstein, R.S.: Fed. Proc. <u>33</u>, 2009 (1974).
Davey, M.G., Lüscher, E.F.: Nature (Lond.) <u>216</u>, 857-858 (1967).
Feinman, R.D., Detwiler, T.C.: Nature (Lond.) <u>249</u>, 172-173 (1974).
Gabbiani, G., Ryan, G.B., Lamelin, J.-P., Vassalli, P., Majno, G., Bouvier, C.A., Cruchaud, A., Lüscher, E.F.: Amer. J. Path. <u>72</u>, 473-484 (1973).
Grette, K.: Norwegian Monographs on Med. Science, 1962.
Grette, K.: Nature (Lond.) <u>198</u>, 488-489 (1963).
Haslam, R.J.: In: Biochemistry and Pharmacology of Blood Platelets. Ciba Foundation Symposium, in press (1975).
Hoffmann-Berling, H.: Ergebn. Physiol. <u>51</u>, 98-130 (1961).
Holmsen, H., Day, H.J., Stormorken, H.: Scand. J. Haemat., Suppl. <u>8</u>, (1969).
Käser-Glanzmann, R., Jakábová, M., Lüscher, E.F.: Some properties of tubulin from human blood platelets. In press (1975).
Kendrick-Jones, J.: Trans. R. Soc. B. <u>265</u>, 183-189 (1973).
Lazarides, E., Weber, K.: Proc. Nat. Acad. USA <u>71</u>, 2268-2272 (1974).
Leger, J., Focant, B.: Biochim. Biophys. Acta <u>328</u>, 166-172 (1973).
Lüscher, E.F., Bettex-Galland, M.: Path. Biol. <u>20</u>, Suppl. 89-101 (1972).
Lüscher, E.F., Massini, P.: In: Biochemistry and Pharmacology of Blood Platelets. Ciba Foundation Symposium, in press (1975).
Lüscher, E.F., Pfueller, S.L., Massini, P.: Ser. Haemat. <u>6</u>, 382-391 (1973).

Lüscher, E.F., Probst, E., Bettex-Galland, M.: Ann. N.Y. Acad. Sci. 201, 122-130 (1972).
Marquis, N.R., Becker, J.A., Vigdahl, R.L.: Biochem. Biophys. Res. Comm. 39, 783-789 (1970).
Massini, P., Lüscher, E.F.: Biochim. Biophys. Acta 372, 109-121 (1974).
Niederman, R., Pollard, T.D.: J. Cell Biol., in press (1975).
Phillips, D.R., Agin, P.P.: Biochim. Biophys. Acta 352, 218-227 (1974).
Pollard, T.D.: In: Molecules and Cell Movement (eds. S. Inone and R.E. Stephens). New York: Raven Press 1975.
Probst, E., Lüscher, E.F.: Biochim. Biophys. Acta 278, 577-584 (1972).
Puszkin, E., Aledort, L., Puszkin, S.: Blood 34, 526 (1969) (Abstr.).
Rosenbluth, J.: J. Cell Biol. 50, 900-1004 (1971).
Salzman, E.W., Chambers, D.A., Neri, L.L.: Nature (Lond.) 210, 167-169 (1966).
Salzman, E.W., Levine, L.: J. Clin. Invest. 50, 131-141 (1971).
Shainoff, J.R.: Ser. Haemat. 6, 392-402 (1973).
Shepro, D., Chao, F.C., Belamarich, F.A.: J. Cell Biol. 43, 129 (1969).
Sokal, G.: Plaquettes sanguines et structure du caillot. Arscia, Bruxelles 1960.
Statland, B.E., Heagan, B.M., White, J.G.: Nature (Lond.) 223, 521-522 (1969).
Taylor, D.G., Mapp, R.J., Crawford, N.: Biochem. Soc. Trans. 3, 161-164 (1975).
Thorens, S., Schaub, M.C., Lüscher, E.F.: Experientia 29, 349-351 (1973).
Weisenberg, R.: Science 177, 1104-1105 (1972).
White, J.G.: In: Platelet Aggregation (ed. J. Caen), p. 16ff. Paris: Masson 1971.
White, J.G.: Amer. J. Path. 66, 295-312 (1972).
White, J.G., Krivit, W.: J. Lab. Clin. Med. 49, 60 (1967).
Wu Loewenhaupt, R., Miller, M.A., Glueck, H.I.: Thrombosis Res. 3, 477-486 (1973).
Zucker-Franklin, D.: J. Clin. Invest. 48, 165-175 (1969).
Zucker-Franklin, D.: J. Cell Biol. 47, 293-299 (1970).
Zucker-Franklin, D., Grusky, G.: J. Clin. Invest. 51, 419-430 (1972).

Discussion

Dr. Th. Wieland: Can you tell us a little more about the two sorts of actin which are present in platelets. Does one know why one of them will not polymerize to give F-actin?

Dr. Lüscher: At the moment we really do not know much about the differences between the two types of molecules. The possibility must be seriously considered that the inability to polymerize is due to damage by handling in the course of the isolation procedure. The two forms would then reflect a particularly vulnerable structure. Amino acid analysis and sequencing probably will give the final answer to this question. It is noteworthy, though, that other authors have confirmed our findings, and more recently, two types of actin have also been described in fibroblasts.

Dr. Th. Wieland: Is it possible to decorate muscle actin with HMM of platelet myosin?

Dr. Lüscher: Yes, indeed, there is cross-reaction.

Dr. Wallenfels: In your list of platelets-aggregating factors you did not mention prostaglandin or prostaglandin peroxide. Why not?

Dr. Lüscher: The listed substances were external agents, which upon addition to platelets will induce aggregation. Prostaglandin E_2 or its precursor endo-peroxide is involved in platelet aggregation, because interruption of its synthesis in the platelet, e.g. by acetyl-salicylic acid, leads to a pronounced impairment of platelet reactivity. PGE_2 is a product of the activated platelet's synthetic activity and should not be confused with an external aggregating agent.

It is perhaps appropriate to mention that other prostaglandins have strikingly different effects on platelets. Thus, PGE_1 is a powerful inhibitor of platelet activity and this is related to its activating effect on adenyl cyclase. All measures taken to increase intracellular cAMP result in an inhibition of platelet activity.

Dr. Podolsky: Is the analogy between platelet aggregation and contractility due to the fact that strips are formed when the platelets aggregate?

Dr. Lüscher: Aggregation is the manifestation of altered surface properties of the platelet. It can be argued that contractile activity is the result of a primary membrane alteration which at the same time predisposes to aggregation. On the other hand, one could equally well postulate that the activation of the contractile system leads to morphological changes - such as the formation of long, thin protrusions in the course of "rapid shape change" - and that it is this "new" surface which is structured in such a way that aggregation becomes possible. I would not even exclude the possibility that an early phase of aggregation, which remains reversible, is simply due to the interlinkage of these long spikes; however, later, so-called second-phase aggregation certainly involves drastic alterations of the membrane structure.

Dr. R. Lamed: Could you comment a little more on the activation of platelets actomyosin by phosphorylation of the myosin component?

Dr. Lüscher: This work has been done by Adelstein. Some of his observation will be published in a forthcoming issue of the journal Haemostasis.

Dr. U. Stewart: In contrast to your finding, we observe a cross-reactivity between antibodies directed against smooth muscle and the myosin from blood platelets. We prepared an antibody against highly-purified myosin from chicken smooth muscle. It has no actin, no troponin or tropomyosin in it but the difference between your antibody and my own is that ours is directed against the globular and the biologically active end of the molecule. Do you think the difference between your and our finding may be dependent on this fact?

Dr. Lüscher: I really don't know. I have been referring to the work of Adelstein (personal communication), who also immunized against smooth muscle myosin. It is difficult to say whether it is the myosin preparation or the mode of immunization which determines the specificity of the antibody. Furthermore, certain individual animals may produce preferably antibodies against one antigenic determinant.

Dr. U. Stewart: I think the mode of immunization does play a role in the immunological determinant of a molecule. We have found that the antibodies against myosin are directed against the head, in at least 50 rabbits over many years, not only in one single animal. The purity of our myosin was checked by SDS-gel electrophoresis and it was at least as pure as Adelstein's material.

Dr. Lüscher: These are interesting findings which certainly should
be followed up.

Dr. Perry: Perhaps I could comment briefly on the phosphorylation of
platelet myosin. I discussed this matter with Dr. Adelstein recently
and it seems that there are some differences between the muscle and
the platelet enzymes. In the first place the muscle enzyme requires
Ca for activity but the platelet enzyme does not appear to require
the cation. Also, even when the platelet actomyosin is fully activated,
it has only about 10% of the activity of normal muscle actomyosin preparation. I think also the enzyme Dr. Adelstein is working with is a
less active preparation than we have for the muscle enzyme. This is
presumably a question of the degree of purification.

Dr. Bryan: If you treat with colchicin or vinblastin, the antimitotic
substances, does the band of microtubules disappear, and if so does
the platelet then form "spikes"?

Dr. Lüscher: Microtubules must in fact disappear, before the disc-
shaped platelet can assume the "spiny-sphere" shape. Their disappearance
most likely is due to the availability of Ca^{2+}-ions within the cell
and therefore coincides with the onset of contractile, or, more specifically, motile activity.

Dr. Bryan: Let me be more specific. If you dissolve the tubules band
for example with colchicin, do you get spiking? I am curious for the
following reason: In tissue cultures which have ruffles which are somewhat analoguous to the spike - I mean at least they both have the filamentous structure in them in tissue culture cells - they tend to be
polarized, i.e. they have ruffles at either end. If you treat with
colchicin and break down the tubules what happens is that the ruffles
are spread all over the cell and they are no longer able to move. It
seems to me that there is some kind of inhibitory interaction going
on between the tubules and the filaments. So if you take the tubules
away intentionally, does then the spiking occur?

Dr. Lüscher: Platelets treated with colchicin or vinblastin, although
morphologically altered, behave quite normally. They aggregate and
show a release reaction. Thus, if spike-formation is an essential part
of platelet activity, then one would conclude that it certainly is not
affected by the disappearance of the microtubules; whether it is increased is unknown to me. Rapid shape change as a rule is investigated
by observing the disc-sphere transformation. Colchicin-treated platelets
are always spheres, so that this method becomes meaningless. To my
knowledge, spike formation under the influence of colchicin has not
yet been studied.

Cytochalasin B, which interferes directly with the contractile system,
has more profound effects on platelet activity. In particular, pseudopode formation is significantly diminished, but interestingly enough,
the release reaction is still observed.

Studies on the Interaction of the Nerve Growth Factor with Tubulin and Actin

P. Calissano, A. Levi, S. Alemà, J. S. Chen, and R. Levi-Montalcini

A. Introduction

The following abbreviations have been used:
NGF = nerve growth factor; MT = microtubule; MF = microfilament; NF = neurofilament; MTOC = microtubule organizing center

Fibrillar proteins can be defined in operational terms as "modular elements" endowed with the property of forming structures which may serve different functions in different cells or also in the same cell during its life cycle. According to this view, the function of micro-filaments (MF) and microtubules (MT) does not only depend on their intrinsic properties but also on the cell-specific activity and re-quirements.

Our interest in fibrous proteins and their role in nerve cell function, was raised by the discovery that the nerve growth factor (NGF), a pro-tein molecule which is a normal constituent of tissue and animal fluids (Levi-Montalcini, 1966), induces the formation of an extraordinarily large number of filamentous structures within the cell body (Levi-Mon-talcini et al., 1968) and enhances the production of nerve fibers in embryonic sensory and embryonic and fully differentiated sympathetic nerve cells (Figs. 1, 2, 3; Levi-Montalcini and Booker, 1960).

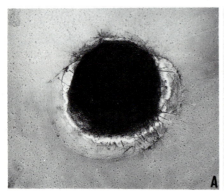 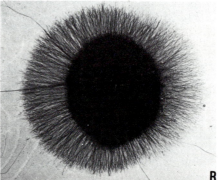

Fig. 1 A and B. Microphotos of sensory ganglia of 8-day chick ganglia cultured *in vitro* for 24 h in semisolid medium without NGF (A), and with the addition of 0.01 µg of NGF (B)

The experiments to be reported here were prompted by an almost casual observation made during a study on the interaction of the NGF with the soluble proteins of the brain (Calissano and Cozzari, 1974). It was

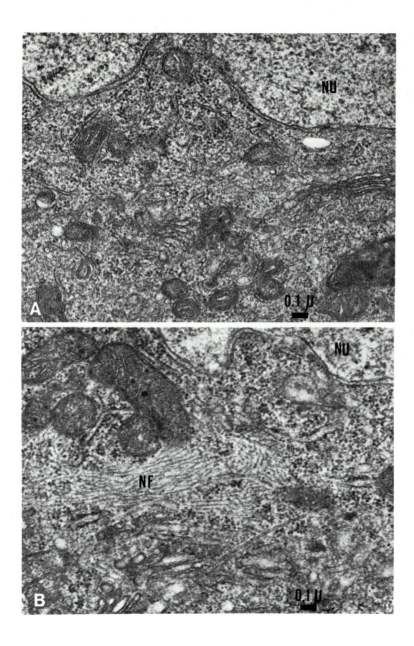

Fig. 2 A and B. Ultrastructural analysis of the short-term action of NGF on sensory ganglia explanted *in vitro*. (A) Ganglia cultured *in vitro* for two hours in absence of NGF (control). Note the almost complete absence of fibrillar elements in the part of cytoplasm close to the nucleus (NU). (B) An analogous electron micrograph of a sensory ganglia incubated for 2 h with 0.01 µg NGF/ml. Note the presence of bundles of organized structures adjacent to the nuclear membrane (NU) and identifiable as microtubules and "filaments" (NF). × 33.200

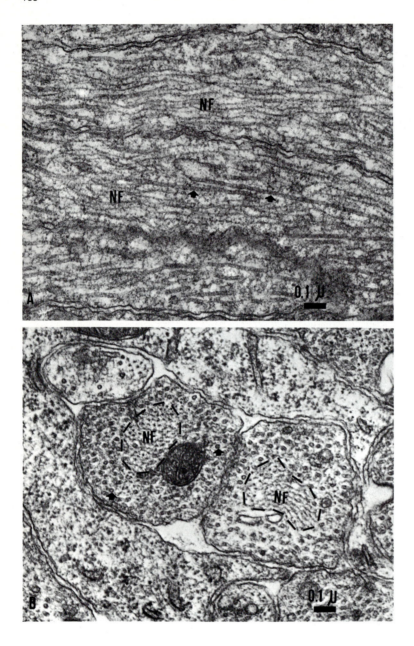

Fig. 3 A and B. Ultrastructural analysis of the "long-term" (24 h) action of NGF on sensory ganglia. (A) Portion of a dorsal root ganglio-cell axon seen longitudinally after incubation for 24 h with 2 μg/ml NGF. Note the filamentous network lining parallel to the long axis of the axon and made of microtubules and other filamentous structures probably formed by neurofilament (NF). × 28.800. (B) Transverse section of the fiber of some ganglion treated with NGF for 24 h. Each axon contains bundles of microtubules and neurofilaments disposed in ordered arrays. × 28.800

found that when NGF is added to the 100,000 x g supernatant of mouse
brain homogenate, an instantaneous aggregation of some protein(s) occurs. This aggregate, separated from the supernatant by centrifugation
and analysed on SDS-acrylamide gel, was found to consist mainly of a
protein identified as tubulin (Fig. 4). Subsequent experiments demonstrated that, in given experimental conditions, other proteins co-pre-
cipitated with tubulin; among these, a component endowed with actin-
like properties. Traces of other high molecular weight proteins are
also found in the pellet (Calissano and Cozzari, 1974).

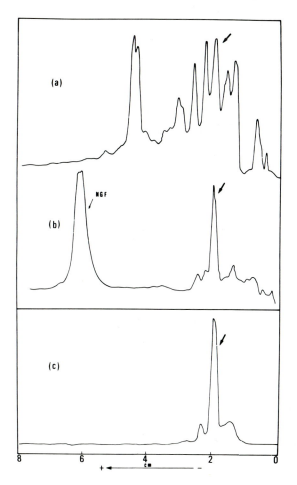

Fig. 4 a-c. Sodium dodecyl
sulfate (SDS) electrophoresis
after nerve growth factor
(NGF) treatment of a 105,000
× g supernatant of mouse brain.
Mouse brain homogenized with
a 1:1 v/v of 0.24 M sucrose
containing 10 mM KH$_2$PO$_4$ pH 6.5
and 10 mM MgCl$_2$ centrifuged at
105,000 × g for 60 min. To
0.2 ml of the supernatant after centrifugation, 1.0 mg of
NGF was added. The fast precipitate formed was allowed
to stand at room temperature
for 10 min, and then centrifuged at 105,000 × g for 30
min. The pellet was dissolved
with 0.1 ml of SDS for electrophoresis, and aliquots were
used for electrophoresis, (a)
= 105,000 × g supernatant before NGF; (b) = 105,000 × g
pellet after NGF addition and
centrifugation; (c) = 50 µg
of purified tubulin. (From
Calissano and Cozzari, 1974)

It was object of the present investigation to examine the nature and
the stoichiometry of binding of NGF with tubulin and actin, the factor
which affect this process, and the possible effects of NGF on (a) the
polymerization of these proteins to form MTs and MFs and, (b) the NGF
action on fibrillar elements preformed *in vitro*. Since muscle actin is
structurally very similar to actin-like protein from the brain (Fine
and Bray, 1971) and is available in large amounts, we made use of muscle actin in the study of NGF-actin interaction.

I. Binding of NGF to Tubulin and Actin

In order to study in detail the interaction of NGF with these proteins, the first important point was the choice of a binding method suitable for the planned studies. The ^{125}I-NGF-tubulin binding assay used in previous experiments (Calissano and Cozzari, 1974) accomplished by precipitation of the complex with vinblastine or by high speed centrifugation, was replaced by two other procedures based on the precipitation of the complexes with ammonium sulfate or on gel filtration. Vinblastine, as a precipitant of the tubulin-NGF complex, was discarded in view of the demonstration that this vinca alkaloid may compete for GTP site(s) on tubulin and that the extent and size of the aggregate formed with tubulin can vary according to the ionic environment (Marantz et al., 1969; Weisenberg et al., 1970; Berry et al., 1972). On the other hand, high speed centrifugation (100,000 × g) of the tubulin or actin-NGF complexes allows the measurement of only the large polymers in the pellet, leaving in the supernatant those complexes small enough to resist high gravity fields. The finding that addition of ammonium sulfate at a final concentration of 35 - 40%, followed by centrifugation, precipitates both the large as well as the small soluble complexes (Levi et al., 1975) made a more precise method of the total binding available. An alternative time-consuming procedure is to chromatograph, on a Sephadex G-75 column equilibrated with ^{125}I-NGF, a small aliquot of tubulin or actin (1 - 2 µg) which, at variance with the growth factor, elutes with the void volume of the column. The radioactivity eluting with tubulin or actin gives the amount of bound NGF.

Fig. 5 shows the binding of NGF to brain tubulin and to chicken muscle actin purified according to Spudich and Watt (1971) at different NGF concentrations. As can be seen, the binding pattern of the two proteins is different, although the concentration of NGF required for saturation is within the same range.

At saturating concentrations, there is an average of 1 mole of NGF bound (m.w. 28,000) per G-actin (m.w. 42,000) and two moles of the growth factor bound per tubulin dimer (m.w. 110,000).

Although we may presume that electrostatic forces play some role in the binding (since tubulin and actin are acidic proteins while NGF has an isoelectric point of 9.3) they do not seem to be essential for the NGF-tubulin interaction since it occurs to a similar extent at 6.0 or 150 mM NaCl. Moreover, the effect of Na$^+$ is to favour, in certain experimental conditions, the interaction of actin and tubulin with NGF, and to change the pattern of binding to the latter from a slightly sigmoid-shaped to an iperbolic curve.

While the actin-NGF curve is typical of 1:1 ratio between the two proteins, the interaction with tubulin shows a somewhat cooperative pattern when Na$^+$ is absent. The effect of Na$^+$, as well as of GTP, is better evidenced when binding studies are performed, instead of with a population of tubulin molecules in different conformational states, with a homogeneous population of dimers, as reported in the following section.

II. Binding of NGF to Tubulin Dimers and Rings or Spirals

Recent investigations have shown that solutions of purified tubulin after centrifugation at 100,000 × g contain this protein as well as traces of other high m.w. proteins in two different conformational states: some as a dimer of 110,000 daltons, and some in the form of

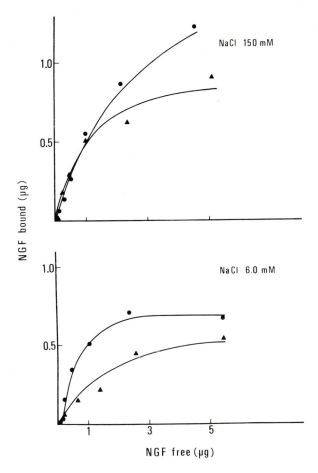

Fig. 5. Binding studies of NGF to tubulin and actin. Binding of ^{125}I-NGF plus various concentrations of unlabelled NGF was studied in the presence (top) and absence (bottom) of 150 mM NaCl in 10 mM MES buffer pH 6.5 containing 1.0 mg/ml of bovine serum/albumin. Purified tubulin (•) (Shelanski et al., 1973) was present in each test tube at a concentration of 1.20 µg, and purified muscle actin (▲) (Spudich, 1971) was 1.15 µg/tube in a final volume of 0.4 ml. After incubation for 60 min at room temperature, an equal volume of 75% $(NH_4)_2SO_4$ was added, tubes were further incubated for 60 min at 8°C and then centrifuged at 6,000 r p.m. in a Sorvall GSA rotor at 2°C for 45 min. After centrifugation, counts in the pellet and supernatant were measured in a Wallac gamma counter

complex structures generally referred to as spirals or rings (Kirschner et al., 1974). These two tubulin populations can be separated by chromatography on agarose 1.5 m columns (Kirschner et al., 1974). With this method, rings and spirals appear in the void volume, while the dimers are retarded by the column. While the former form MTs when incubated in suitable conditions, the latter is not able to self-assemble unless mixed with a portion of rings. We then measured the binding of NGF to these two forms of tubulin molecules in the presence of substances (GTP, Na+) which could affect the conformation of tubulin and/or its organization. Fig. 6a shows that while NGF binds to the tubulin dimers in a markedly cooperative fashion, its interaction with the same molecule organized in the form of rings and spirals follows an almost iperbolic course. Stronger evidence that NGF interacts with tubulin dimers in a cooperative manner was obtained by means of column chromatography, which measures binding at equilibrium and does not require addition of salts like $(NH_4)_2SO_4$, which could interfere with the total binding. With this procedure also, binding exhibits a typical sigmoid pattern. NaCl affects the dimers in such a way that their ability to interact with NGF is almost identical with that exhibited by the rings, i.e. the monovalent cation changes a cooperative interaction into a first-order reaction as shown in Fig. 6b. GTP produces an analogous effect: in its presence the interaction of NGF with tubulin dimers follows an iperbolic course identical to that obtained with rings and spirals (Fig. 6c). These results suggest that both NaCl and GTP induce a change in the conformation or in the quaternary structure of tubulin dimers in such a way that they are transformed in structures behaving like rings or spirals when interacting with NGF.

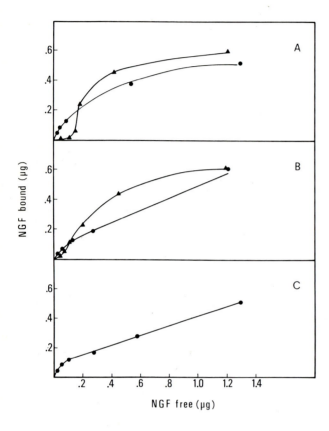

Fig. 6 A-C. Binding of ^{125}I-NGF to rings and dimers of tubulin in presence of NaCl or GTP. Tubulin dimers (A) and spirals and rings (B), were separated by column chromatography (Kirschner et al., 1974) and assayed for their binding ability of NGF in the presence (●) and absence (▲) of 150 mM NaCl. (C) Analogous experiment performed with tubulin dimers in the presence of 1.0 mM GTP. Tubulin dimers were 1.2 μg/test tube and rings and spirals 1.6 μg. All procedures as in Fig. 5

III. Correlation of NGF Binding with Tubulin Aggregation

As shown in Fig. 4, the interaction of NGF with tubulin results, under some experimental conditions, in the formation of large complexes which confer turbidity on the protein solution and sediments under low gravity acceleration. We attempted to correlate the binding of NGF with the induced aggregation and to follow the effect of substances (Na^+, GTP, H^+) which do not affect the total binding, but could alter the dimensions of the complexes formed by the two proteins. Fig. 7 shows that maximum binding occurs at pH 7.4, while formation of large complexes, measurable as increase in light scattering, is accomplished within the optimal pH of 6.0 - 6.5.

Fig. 8 shows the stoichiometry of binding as correlated with the formation of large complexes. It can be seen that the formation of these large NGF-tubulin complexes exhibits a markedly cooperative pattern and that the increase in turbidity (Fig. 8a) follows a similar if not identical binding pattern of NGF (Fig. 8b) and tubulin (Fig. 8c) indicating that the binding and induced aggregation are strictly related phenomena. The molar ratio of NGF/tubulin in the precipitates varies

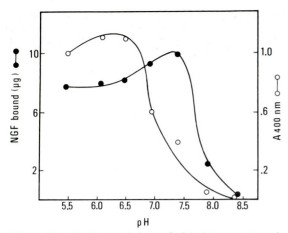

Fig. 7. Binding (●) and light scattering (○) of the tubulin-NGF complexes at different H+ concentrations. The concentrations of tubulin
and NGF were 45 and 35 μg in a final volume of O.25 ml of RB at different pHs. Na+ concentration was maintained constant at every pH by
adding the proper amount of NaCl. After 15 min at room temperature,
the absorbancy of the solutions was read and immediately afterwards
an equal volume of 70% $(NH_4)_2SO_4$ was added to measure the amount of
NGF bound to tubulin. (From Levi et al., 1975)

between 1:1 and 2-3:1. 1.O mM GTP shifts the steepest part of the
curves, both of light scattering and binding, so that more NGF is
needed to trigger the aggregation; this is true also for NaCl, whose
effect at 150 mM concentration is more pronounced than that of GTP
itself. Since previous experiments had shown that these two substances
do not significantly interfere with the binding, they must affect the
organization and/or dimension of the aggregates. Experiments where
turbidity was monitored in the presence of different concentration of
tubulin (Fig. 9) demonstrated that there is a linear relationship between the concentration of tubulin present in solution and the amount
of NGF necessary to trigger the sudden increase in light absorbance.
In other words, a molar ratio of NGF/tubulin of at least 1:1 is required to bring about the formation of large complexes. This is not
the case for actin, which starts aggregating at the same concentration
of NGF, independently from its amount in the incubation mixture (Alema,
unpublished data).

In summary, the interaction between NGF and tubulin, analyzed by different methods, reveals peculiar properties which seem to be of a
specific and significant nature. The binding is in fact stoichiometrically precise, since each subunit forming the tubulin dimer shows the
capacity of binding one NGF molecule. It is possible that the different
primary structure of the alpha and beta subunits of tubulin is responsible for the non-linear binding of NGF, which could have different
affinities for them and/or it could behave as if the binding of the
first molecule facilitates the second to interact with the other subunit. This process can be modulated by NaCl, GTP and probably by divalent cations (Calissano and Cozzari, 1974). Thus, these substances
seem to play a double control on NGF-tubulin interaction. First, by
regulating intramolecular forces which allow the binding between the
two molecules, and then by regulating the size and shape of the supramolecular structure of these complexes, probably by inhibiting ionic
interactions between 1:1 or 2:1 complexes (see Fig. 8).

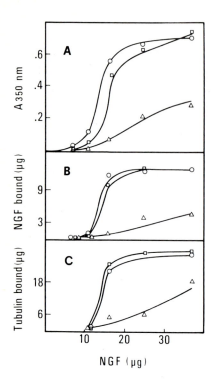

Fig. 8 A-C. Light scattering and formation of the "heavy structures of NGF-tubulin complexes. A constant amount of tubulin (84 µg) was added to tubes containing different concentrations of NGF in a final volume of 0.75 ml of 10 mM MES, pH 6.5 0.5 mM MgCl₂ and 1.0 mM EGTA. After 15 min at room temperature, 3 aliquots of o.25 ml were taken for each NGF concentration and 150 mM NaCl (△), 1.0 mM GTP (□) or the corresponding volume of buffer (○) was added to each set of tubes, the solutions mixed and allowed to stand for 10 min. After reading the light scattering at 350 nm (A) the solutions were centrifuged twice, 2 min each at 10,000 rpm in a Beckman 152 microfuge. NGF (B) and tubulin (C) precipitated, were measured both as appearance in the pellet and disappearance in the supernatant with the use of ¹²⁵NGF and the colchicine binding assay (17) respectively. (From Levi et al., 1975)

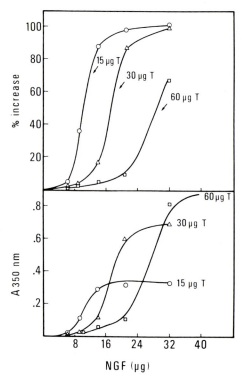

Fig. 9. Light scattering at three different concentrations of tubulin: 15 µg (○); 30 µg (△); 60 µg (□) in a volume of 0.3 ml 10 mM MES + 0.5 mM MgCl₂ + 1 mM EGTA. After the addition of NGF at the concentration indicated in the abscissa, the solutions were allowed to stand at room temperature and the absorbancy read at 350 nM. Lower half of figure: absolute absorbancy measurements; upper half: % increase of absorbancy calculated as

$$\frac{A}{A\ max} \times 100. \text{ (From Levi et al.,}$$

1975)

IV. Effect of NGF on Tubulin Polymerization

Assembling of tubulin *in vitro* was described at the E.M. at a suitable
pH (6.5) and in presence of GTP, Mg^+ and Ca^{++} chelators (Weisenberg,
1972). Quantitative data on this process were obtained by viscosity
(Olmsted and Borisy, 1973) and turbidity determinations (Shelanski
et al., 1973). The latter provides a simple and reliable test of tubulin polymerization and consists in measuring an increase in light
scattering (visible light 350 - 400 nm) of a solution of tubulin incubated at 25 - 37°C.

Fig. 10 shows that NGF, added at a concentration 25 - 30 times lower
than tubulin, accelerates the rate of assembly to form MTs. This effect is obtained even after several cycles of polymerization - depolymerization, but does not occur when substances such as Ca^{++}, colchicine or a low H^+ concentration (pH 7.5 instead of 6.5) are present in
the incubation medium.

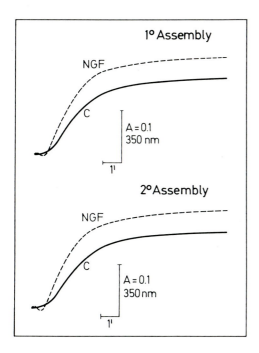

Fig. 10. Rate of microtubule formation in absence and presence of
NGF. Rate monitored as turbidity
increase in termostated cuvettes
(30°C) with tubulin at a concentration of 2.0 mg/ml. NGF present
at a concentration of 0.1 mg/ml.
After incubation for 10 min at
30°C in absence (——) or in presence (---) of NGF, 1.0 mM GTP
was added and the increase in
light scattering measured. The
figure shows two cycles of assembly. Disassembly, after the first
polymerization, was obtained by
cooling on ice the solution of
tubulin ± NGF. (From Levi et al.,
1975)

The finding that NGF accelerates the rate of polymerization and the
apparent net amount of MTs and at the same time enhances the formation
of organized tubulin structures suggests that it could act, to some
extent as the "nucleation centers" which catalyze *in vitro* the polymerization of tubulin (Borisy and Olmsted, 1972). This protein is not able
to self-assemble when these nucleation centers are separated by column
chromatography, high speed centrifugation or when the tubulin is diluted below its critical concentration (0.1 - 0.2 mg/ml) (Kirschner
et al., 1974). Incubation of NGF with diluted tubulin solution (NGF-
tubulin molar ratio = 1.0) results in the formation of MTs as shown
in Fig. 11. These structures appear of normal length and diameter,
although their number is much smaller than that obtained by an identical solution of tubulin to which the nucleation centers have been

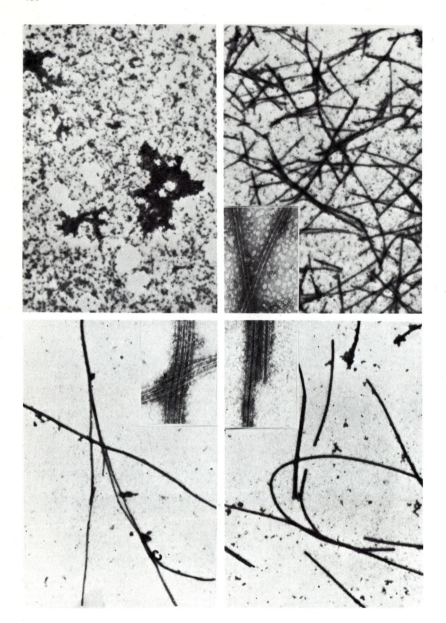

Fig. 11. Effect of NGF and lysozime on microtubule formation. The for-mation of microtubule induced by NGF or lysozime was obtained by in-cubating at 37°C for 30 min a solution of tubulin at different concen-trations in 100 mM MES, 0.5 mM MgCl$_2$, 1.0 mM EGTA and 1.0 mM GTP. Upper left: tubulin 0.15 mg/ml. Upper right: tubulin 1.0 mg/ml. Lower left: tubulin 0.15 mg/ml + NGF 0.15 mg/ml. Lower right: tubulin 0.15 mg/ml + lysozime 0.15 mg/ml. All photographs were taken at 18,000 ×; the insets show a higher magnification view (56,000 ×) of the same preparation. Negative staining with uranil acetate 1%. (From Levi et al., 1975)

added. The cause of this difference may be due to the fact that the incubation medium and experimental conditions are still not optimal for the two species of molecules. In evaluating this NGF effect, it is important to keep in mind that some other basic proteins (alkaline phosphatase, histones, and polysine) also induce precipitation of tubulin and affect its polymerization. Among these, lisozyme induces an effect somewhat similar to NGF (Fig. 11), although the tubules have a diameter 10 - 20% larger than normal.

V. Interaction of NGF with Actin

The iperbolic curve of NGF binding to muscle G-actin has been shown in Fig. 5. This type of curve together with a stoichiometry of apparently 1:1 at saturation indicates that interaction between the two molecules is a kinetically simpler reaction than that which takes place between NGF and tubulin, even if the association constant must be at least equally high. It was found that when actin is allowed to polymerize to form MFs (F-actin) and the NGF-actin binding is studied by the ammonium sulphate method, the binding to F-actin is 20 - 25% of that of G-actin under the same experimental conditions.

Fig. 12 shows the effect of NGF on preformed microfilaments. It can be seen that when NGF is added at a final concentration of 0.1 - 0.2 moles/ mole of F-actin, the distribution of MFs which is normally random becomes linearly oriented. An increase in concentration of NGF to 1.0 mole/mole F-actin, results in the formation of "fibers" which, at higher magnification, appear to consist of several MFs ordered in multiple arrays up to 7 - 10 in number. Each of these fibers exhibits a transverse striation (Fig. 12) spaced at approximately 400 Å intervals, resembling the paracrystals induced by high Mg^{++} concentration (Spudich, 1973). It is worth mentioning that very similar, although nonidentical organizations of MFs have been described in cells such as fibroblasts (Goldman et al., 1975) and in Limulus sperm (Tilney, 1975). When NGF is added to a solution of (G-actin) it induces the formation of large complexes which scatter a light beam at 350 nm in a way similar to NGF-bound tubulin. These aggregates do not show a well-defined organization in the absence of KCl. 100 mM KCl and 2.0 mM $MgCl_2$ bring about a rearrangement of NGF-actin complexes to form fibers similar to those produced by addition of NGF to preformed MFs. These findings suggest that the NGF site of actin is distant from those responsible for monomer interactions in MFs formation. The alternative possibility that NGF might be progressively displaced from actin during its assembly is less likely, although it cannot be definitely ruled out, since binding of NGF to preformed MFs is still measurable, even if reduced to 20 - 25%.

In evaluating the NGF effect on MFs it is of some interest to mention that phalloidin, a byciclic peptide which produces intense degenerative effects in liver cells, has been recently shown to induce irreversible polymerization of intracellular actin (Lengsfeld, 1974; Low et al., 1974). Its *in vitro* effect consists in favouring formation of MFs in the absence of KCl and in stabilizing preformed MFs against the disintegrating action of 0.6 M KI. Neither one of the above effects was obtained with NGF.

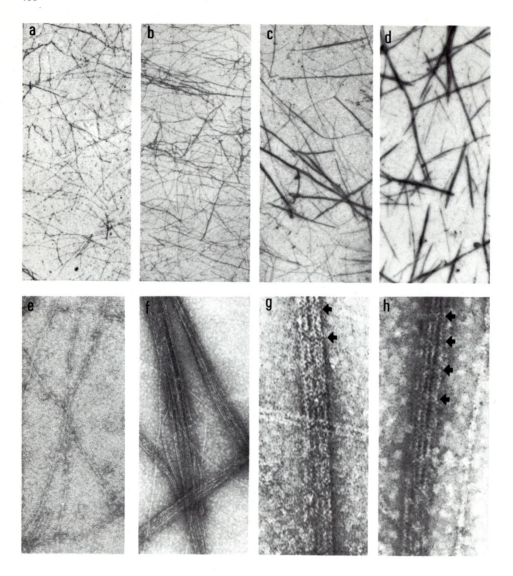

Fig. 12 a-h. Electron micrographs of NGF added to preformed F-actin filaments. Solutions of purified muscle actin 0.5 mg/ml (Spudich, 1971) were incubated in 5.0 mM Tris-Cl buffer pH 8.1 containing 100 mM KCl, 2.0 mM $MgCl_2$ and 2.0 mM ATP for 30 min. After polymerization, an aliquot was directly stained with 1% uranil acetate (a) while the remainder was divided into 0.2 ml portions to which 12 (b), 30 (c) and 60 (d) μg of NGF were added. After incubation for 30 min at room temperature they were stained with 1% uranil acetate. Magnification of (a-d) = × 8,500; (e) = × 84,800 and (f) = × 63,600 show higher magnification of samples (a and d). (g and h) represent a × 180,000 magnification of (c and d) showing the transverse striation (arrows)

B. Discussion and Concluding Remarks

It was the object of this investigation to study the NGF-tubulin and NGF-actin interaction and the kinetics of both processes under different experimental conditions. The results reported in previous sections bring to light some new aspects of these interactions which appear to be highly specific and raise the question of their significance and of the primary or secondary role which they may play in the chain of events triggered by the nerve-growth factor in its target cells.

It was already mentioned that one of the earliest if not the earliest detectable effect in embryonic sensory and sympathetic nerve cells upon *in vitro* or *in vivo* NGF treatmet, is the production of an extraordinarily large number of neurotubules and neurofilaments which fill the cell perikaria and their axons.

Studies on the mechanism of action of NGF uncovered other aspects of this stimulus-response system which call attention to fibrillar proteins and their possible implication in mediating NGF activity on its receptive nerve cells. These are: (a) the evidence of specific NGF receptors on the plasma membranes of embryonic sensory (Herrup and Shooter, 1973) sympathetic (Frazier et al., 1974; Banerjee et al., 1973) and neuroblastoma cells (Revoltella et al., 1974) and (b) the demonstration that the NGF molecule is transported in a retrograde fashion from the nerve endings to the body of sympathetic nerve cells (Hendry et al., 1974; Stockel et al., 1974; Iversen et al., 1975). While in the latter case, microtubules and possibly also microfilaments play a well-reconized and essential role in the mechanics of transport processes, in the former only indirect evidence was presented for a possible involvement of fibrillar proteins in mediating NGF action on its target cells (Levi-Montalcini et al., 1974). We shall first consider the NGF-microtubules and NGF-microfilaments interactions in relationship to the retrograde axonal transport of this molecule.

Transport can take place only through NGF binding followed by interiorization and active transfer from the nerve endings to the cell body. While active ans specific NGF retrograde axonal transport in sympathetic nerve fibers has been definitely proved, the molecular mechanism underlying this transport have not yet been elucidated. It is tempting to consider the results reported in this investigation as pertinent to this problem.

The ability of MTs and MFs of binding to NGF, as well as the finding that these processes can be modulated by metabolites and ions which are normal constituents of animal tissues, lend support to the hypothesis that these interactions play a physiological role in mediating the NGF effect on the receptive nerve cells. Viewed in this light, MTs and MFs would act as transport channels, a role greatly facilitated by the selective affinity of NGF for these fibrillar proteins.

Recently, multiple and diversified role have been prospected for these proteins which - as suggested in the introduction - could be conceived of as "modular elements" which can partake in a number of processes according to the cell's special needs and functional properties. It is of particular interest in this connection to mention among the several roles attributed to tubulin, one related to <u>intercellular</u> rather than <u>intracellular</u> communication systems.

Studies on surface receptors and various ligands shows a striking colchicine effect on receptor mobility (Edelman et al., 1971) as well as
on lymphocyte-mediated cytotoxicity (Plaut et al., 1973). Hence the
concept developed that tubulin may be endowed with an all-important
role in membrane surface configuration which would, in turn set in
motion a chain of cytoplasmic and nuclear processes (Levi-Montalcini
et al., 1974). Similar roles were hypothetized for actin-like protein
(Allison, 1972). Both fibrillar proteins have been shown to be associated with plasma membrane isolated from cell perikaria as well as
from nerve endings (Blitz and Fine, 1974).

In the system under consideration, binding of NGF to the receptor or
to tubulin itself on the cell membrane could result in a local concentration of the growth factor sufficient to trigger the formation
of patches of tubulin molecules. These complexes could, in turn, act
as (a) "nucleation centers", (b) sites for attachment to the membrane
of preformed cytoplasmic microtubules, (c) complexes favouring NGF
transport across the membrane, (d) modulators of enzymatic activities
or ionic fluxes.

Similar mechanisms could be postulated for the actin-like filaments
which are mainly confined to the tips of nerve fibers (Yamada et al.,
1971; Bray, 1973).

Studies in progress are directed at gaining more precise information
on the characteristics of the interaction between NGF and fibrillar
proteins and on the sequence of events triggered by NGF binding at
the cell surface and/or interiorization inside the cells.

References

Allison, A.C.: In: Cell Interactions (ed. L. Silvestri), pp. 156-161.
  Amsterdam: North Holland 1972.
Banerjee, S.P., Snyder, S.H., Cuatrecasas, P., Greene, L.A.: Proc.
  Nat. Acad. Sci. 70, 2519-2523 (1973).
Berry, R., Shelanski, M.L.: J. Mol. Biol. 71, 71-80 (1972).
Blitz, A.L., Fine, R.E.: Proc. Nat. Acad. Sci. 71, 4472-4476 (1974).
Borisy, G.G., Olmsted, J.B.: Science 177, 1196-1197 (1972).
Bray, D.: Nature 244, 93-96 (1973).
Calissano, P., Cozzari, C.: Proc. Nat. Acad. Sci, 71, 1231-1235 (1974).
Edelman, G.M., Yahara, I., Wang, J.L.: Proc. Nat. Acad. Sci. 70, 1442-
  1446 (1973).
Fine, R.E., Bray, D.: Nature New Biol. 234, 115-118 (1971).
Frazier, W.A., Boyd, L.F., Bradshaw, R.A.: J. Biol. Chem. 249, 5513-
  5519 (1974).
Gaskin, F., Kramer, S.B., Cantor, C.R., Adelstein, R., Shelanski, M.L.:
  FEBS Letters 40, 281-285 (1974).
Goldman, R.D., Lazarides, E., Pollock, R., Weber, K.: Exp. Cell Res.
  90, 333-344 (1975).
Hendry, A., Stöckel, K., Thoenen, H., Iversen, L.L.: Brain Res. 68,
  103-121 (1974).
Herrup, K., Shooter, E.M.: Proc. Nat. Acad. Sci. 70, 3884-3888 (1973).
Iversen, L.L., Stöckel, K., Thoenen, H.: Brain Res. 88, 37-43 (1975).

Kirschner, M.W., Williams, R.C., Weingarten, M., Gerhart, J.C.: Proc.
  Nat. Acad. Sci. 71, 1159-1163 (1974).
Legsfeld, A.M., Low, I., Wieland, T., Dancker, P., Hasselbach, W.:
  Proc. Nat. Acad. Sci. 71, 2803-2807 (1974).
Levi, A., Cimino, M., Mercanti, D., Chen, J.S., Calissano, P.: Biochem.
  Biophys. Acta 399, 50-60 (1975).
Levi-Montalcini, R.: Harvey Lect. 60, 217-259 (1966).
Levi-Montalcini, R., Booker, B.: Proc. Nat. Acad. Sci. 42, 373-384
  (1960).
Levi-Montalcini, R., Caramia, F., Luse, S.A., Angeletti, P.U.: Brain
  Res. 8, 347-362 (1968).
Levi-Montalcini, R., Revoltella, R., Calissano, P.: Recent Progr. in
  Hormone Res. 30, 635-669 (1974).
Marantz, R., Ventilla, M., Shelanski, M.: Science 165, 498-499 (1969).
Olmsted, J.B., Borisy, G.G.: Biochemistry 12, 4282-4289 (1973).
Paravicini, U., Stöckel, K., Thoenen, H.: Brain Res. 84, 279-291
  (1975).
Plaut, M., Lichtenstein, L.M., Henney, C.S.: J. Immunol. 110, 771-780
  (1973).
Revoltella, R., Bertolini, L., Pediconi, M., Vigneti, E.: J. exp. Med.
  140, 437-451 (1974).
Shelanski, L.M., Gaskin, F., Cantor, C.R.: Proc. Nat. Acad. Sci. 70,
  765-768 (1973).
Spudich, J.: Cold Spring Harbor Symp. Quant. Biol. 37, 585-593 (1973).
Spudich, J.A., Watt, S.: J. Biol. Chem. 246, 4866-4871 (1971).
Stöckel, K., Paravicini, V., Thoenen, H.: Brain Res. 76, 413-421
  (1974).
Tilney, L.G.: J. Cell Biol. 64, 289-310 (1975).
Weisemberg, R.C.: Science 177, 1104-1105 (1972).
Weisemberg, R.C., Timasheff, S.: Biochemistry 9, 4110-4116 (1970).
Yamada, K.M., Spooner, B.S., Wessel, N.K.: J. Cell Biol. 49, 614-636
  (1971).

Discussion

Dr. Hamprecht: As you know, complete NGF consists besides the β-sub-unit also of α- and γ-subunits. Does the complex behave like the β-subunit in your binding studies on tubulin and actin?

Dr. Calissano: We have not done the experiments you are asking about, but we are planning to do them in the near future. I wish to say, how-ever, that as far as I know the α- and γ-subunits do not have any bio-logical effect. Binding studies with the entire 7S complex are thus curtainly interesting but do not appear essential for the possible biological significance of the interaction of the β-subunit with tu-bulin or actin.

Dr. Hamprecht: Do you know whether the β-subunit enters the cytoplasm of the cells?

Dr. Calissano: The evidence that NGF enters the cytoplasm of differentiated sympathetic cells is now well established by the studies which I mentioned on the retrograde axonal transport of this growth factor. On the other hand, we now have some evidence that after incubation of J^{125} NGF with chick embryo sympathetic or sensory cells for a few hours, a portion of the growth factor ranging between 2% and 20% is not digested by the common proteolytic enzymes. This finding encourages one to think that also in these cells NGF can be interiorized, a possibility which is now under further investigation in our laboratory.

Phallotoxins and Microfilaments

Th. Wieland

A. Some Chemistry of the Phallotoxins

Phalloidin (Formula 1a) is one of the toxic components of the toadstool *Amanita phalloides* (Wieland, 1968); it is accompanied in the mushroom by several toxic relatives of which only phallacidin (Formula 1b) is shown in the general formula. Chemical manipulations can also lead to toxic derivatives or can annihilate the toxicity of the molecule. By degradation of the branched side chain 1a is transformed to desmethylphalloin (Formula 1c), a still toxic product, which can also be obtained in the tritiated state by this reaction. Radioactive carbon can be introduced by methylation of the indole nucleus of 1a giving a still toxic N^{ind}-methyl compound (Formula 1d) (Faulstich and Wieland, 1971). On oxidation with hydrogen peroxide in acetic acid, two diastereomeric sulf-

	R_1	R_2	R_3	R_4	toxicity
(a) Phalloidin	C(OH)–CH$_2$OH CH$_3$	CH$_3$	CH$_3$	–S–	+
(b) Phallacidin	like (a)	CO$_2$H	CH(CH$_3$)$_2$	–S–	+
(c) Desmethyl- phalloin	CH(OH)–CH$_3$	CH$_3$	CH$_3$	–S–	+
(d) N^{ind}-Methyl- phalloidin	like (a), but CH$_3$ instead of H at indole-N				+
(e) (R)-Phalloidin- sulphoxid	——	like (a)	———	$-\overset{..}{s}\overset{\nearrow O}{-}$	+
(f) (S)-Phalloidin- sulphoxid	——	like (a)	———	$\overset{O}{\underset{}{-\overset{..}{s}-}}$	–
(g) Dethiophalloidin	——	like (a)	———	–H H–	–
(h) Seco(15)-phalloidin	——	like (a), but peptide bond split (arrow)			–

oxides are formed from 1a (Faulstich et al., 1968), a toxic one (Formula 1e) with R-chirality at the SO center and a nontoxic one (Formula 1f) with S-chirality (Wieland et al., 1974). Vanishing of toxicity is also observed after opening of the thioether bridge by replacing the sulfur by two H-atoms by means of Raney-nickel (dethiophalloidin, Formula 1g) or by selective fission of a predestinated peptide bond which leads to a secophalloidin (Formula 1h).

B. The Symptoms of Phalloidin Poisoning

The symptoms of phalloidin poisoning (Wieland and Wieland, 1972) point to an impairment of the cytoplasma membrane of hepatocytes: after administration of 3 mg of the toxin per kg; the animals (white mice) will die within 1-2 h with livers swollen to as much as double their weights. The swelling is due to an excessive accumulation of blood in the liver; this final stage of acute haemorrhagic necrosis is preceded by the formation of numerous non-fatty vacuoles, which have their origin in an endocytosis, which begins 2.5 min after addition of the poison to a perfused isolated rat liver preparation (Lengsfeld and Jahn, 1974). An analogous uptake of water can also be produced without phalloidin by increasing the posthepatic pressure (Jahn, 1972). So it seemed that phalloidin weakens the membrane in such a way as to take up by endocytosis outcellular liquid already under the pressure normally present in the liver. Since also an efflux of K^+-ions had been observed as a consequence of phalloidin action (Frimmer et al., 1967) and a binding of the toxin to plasma membranes of liver cells had been stated (Lutz et al., 1972), an electron microscopic study was undertaken on membrane fragments obtained by gradient centrifugation from intoxicated rats (Govindan et al., 1972). Numerous filamentous structures about 60 Å in width and up to 1 μm in length were regularly observed, which were

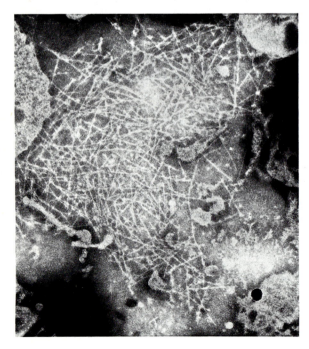

Fig. 1. Electron micrograph of microfilaments produced by incubation of plasma membrane fraction of rat liver with phalloidin. × 60,000. (Photo: A.M. Lengsfeld)

only occasionally present in similar preparations from control animals.
Analogous structures could also be produced *in vitro* by incubation with
phalloidin of cytoplasmic membrane preparations from the livers of un-
poisoned rats. Using ^{3}H-labelled 1c (see Formula 1) we proved that the
toxin was bound mainly to the filaments (Govindan et al., 1973). Later
on it could be demonstrated by decoration with heavy meromyosin (HMM)
that the phallotoxin-induced liver filaments consist of actin (Lengsfeld et al., 1974), which however proved resistant to degradation by
0.6 M KI, a reagent which is known to depolymerize normal F-actin (Fig.
2). This prompted us to start an investigation on the interaction of
actin from rabbit muscle with phallotoxins.

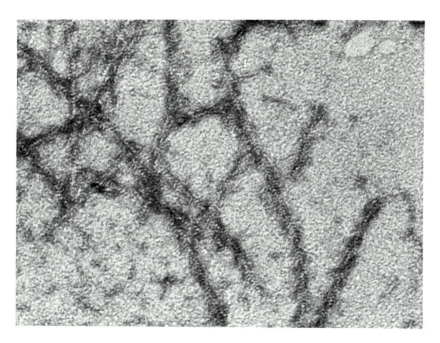

Fig. 2. Electron micrograph of phalloidin-induced liver filaments after
decorating with HMM. Negativ staining with phosphotungstate. × 160,000.
(Photo: A.M. Lengsfeld)

C. Phallotoxins and Actin

The system G-actin $\rightleftarrows$ F-actin (without troponin/tropomyosin) can be influenced in several ways. Polymerization of G-actin needs Ca^{++} or Mg^{++},
KCl and ADP or ATP (the latter being split into $ADP + P_i$ during the
attachment of the G-units). Depolymerization of F-actin occurs after
decreasing the ionic strength, by addition of ATP (which stabilizes the
monomers) or by high concentrations (0.6 M) of KI or KSCN.

$$\text{G-actin} \xrightleftharpoons[\text{ATP, low ionic strength, KI, KSCN}]{Mg^{++}(\text{or } Ca^{++}),\ ADP(\text{or } ATP \rightarrow ADP + P_i),\ K^+Cl^-} \text{F-actin}$$

I. Polymerization of G-actin

By <u>viscosimetry</u> (Löw and Wieland, 1974) it was shown that phalloidin
induces polymerization of G-actin also in the absence of K^+-ions. Polymerization under these conditions is also induced by all toxic and non-
toxic derivatives (Table 1); most strikingly, however, the high viscosity is practically maintained after KI in all cases where toxic

Table 1. Specific viscosities of solutions of G-actin (1.2 mg/ml 1 mM
Tris-HCl, pH 7.4 + 0.7 mM Mg^{++}) 30 min after incubation with different
toxic and non-toxic cyclic peptides (80 µg) before and after addition
of KI (0.6 M)

Compound added	η spec. Before	After added KI
None	O	–
KCl (0.1 M)	1.04	O
Phalloidin (1a)	1.04	0.93
Phallacidin (1b)	1.0	0.70
Desmethylphalloin (1c)	1.2	0.94
(R)Phalloidin-sulfoxide (1e)	1.1	0.94
(S)Phalloidin-sulfoxide (1f)	1.05	O
Dethiophalloidin (1g)	1.0	O
Secophalloidin (1h)	1.0	O

compounds had been added. The stabilization of F-actin against KI by
phalloidin occurs also in the polymerized state: F-actin filaments
can be made resistant by incubation with the toxin. The transformed
F-actin has been named Ph-actin.

The rate of polymerization of G-actin induced by different amounts of
phalloidin has been measured by observing the increase of light-scattering intensity of the actin solution at 400 nm (Dancker et al.,
1975). Polymerization was started by the addition of $MgCl_2$ (to 1 mM)
to 58 nmoles of G-actin in 0.1 mM ATP.

Fig. 3 shows the accelerating effect of the toxin on the rate of polymerization. As little al 5 nmoles (molar ratio toxin/actin 0.09) of
1a distinctly increased the velocity as compared with the control and
further gradual increase gave rise to a gradual increase of the rate.
The maximal rate was reached not before phalloidin and actin were
present in equimolar amounts. Phalloidin even brings about polymerization of G-actin in presence of KI (Fig. 4).

Whereas no polymerization occured without the drug already in presence
of 0.3 M KI, the toxin (2 moles pro mole G-actin) induced Ph-actin
formation even in 0.5 m KI. There exists something like an antagonism,
for with decreasing concentration of iodide an increasing velocity of
polymerization can be observed.

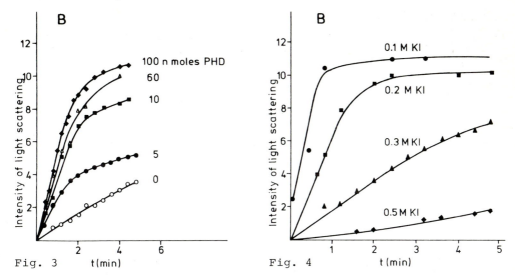

Fig. 3   t(min)   Fig. 4   t (min)

Fig. 3. Rate of polymerization of G-actin (58 nmoles) accelerated by increasing amounts of phalloidin. -o-o- without PHD -♦-♦- with 100 nmoles PHD

Fig. 4. Inhibition of phalloidin-induced polymerization rate of G-actin by KI

II. Stabilization of F-actin

As already mentioned, phallotoxins are able to stabilize F-actin against depolymerization by O.6 M KI (Ph-actin). In order to find out a stoichiometric relation a viscosimetric assay was made of the pro- tective effect of different amounts of the cyclic peptide on F-actin (Dancker et al., 1975). From Fig. 5 it can be seen that F-actin was

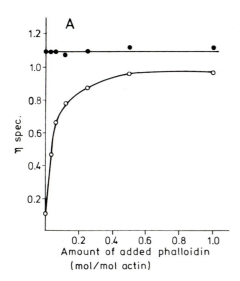

Amount of added phalloidin
(mol/mol actin)

Fig. 5. Drop of viscosity by O.6 M KI of F-actin stabilized with dif- ferent amounts of phalloidin

completely protected only when phalloidin and the actin subunits were present in equimolar amounts, although a high degree of protection was already reached when the molar ratio of phalloidin to actin was very small. Here it may be added that Ph-actin is also stable against 1 mM ATP in ion free medium (Löw and Wieland, 1974).

Next the heat tolerance of actin was studied in the presence of varying amounts of phalloidin (Wieland et al., 1975a). The results are shown in Fig. 6, in which the degree of denaturation by exposing the sample

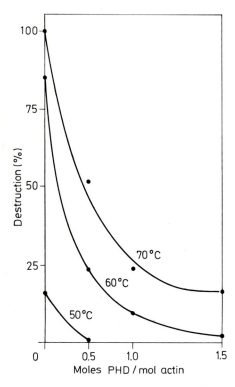

Fig. 6. Protective effect of phalloidin against heat denaturation (3 min) of actin

for 3 min to different temperatures is plotted against increasing amounts of the toxin. As one can see, the protein is destroyed more than 80% at 60°C in the absence of phalloidin but remains almost intact when 2 moles phalloidin per mole of actin are present.

Besides stabilizing against heat denaturation, phalloidin also protects actin from combination with deoxyribonuclease I (Schäfer et al., 1975). This enzyme has been shown to bind very specifically with actin, which accordingly is the long-known naturally occuring inhibitor (Lazarides and Lindberg, 1974). We observed an immediate turbidity on putting together the solutions of both the proteins; after preincubation of the actin with an equimolar amount of phalloidin, however, the solution remained absolutely clear.

F-actin filaments proved also stabilized by phalloidin against sonic vibration (Dancker et al., 1975). Asakura showed as early as 1961 that during ultrasonication F-actin in a ATP-containing solution gives rise to ATP splitting into ADP and inorganic phosphate. Obviously sonic vibration causes local loosenings of the F-actin structure, which allows the exchange of bound ADP with ATP of the medium. The subsequent healing of the local perturbation induces the hydrolysis of ATP in a

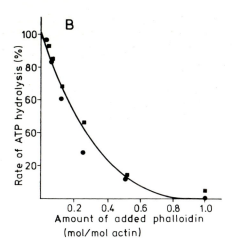

Fig. 7. ATPase induced at F-actin
by ultrasonic vibration is suppressed by increasing amounts of
phalloidin

way reminiscent of the ATP splitting during polymerization of G-actin.
Fig. 7 shows that the ATPase activity of F-actin during sonic vibration
can be completely inhibited by phalloidin. The ATP hydrolysis was already diminished considerably when only a small proportion of the actin
units could have combined with phalloidin. The line drawn in the figure
represents the course of probability for 3 adjacent actin units remaining free, if the total added phalloidin were bound in a purely
statistical manner, each actin unit being one binding site. This would
mean that breakage of the chain can only occur between at least 3
phalloidin-free units.

That phallotoxins are bound to actin has been shown using ^{3}H-desmethyl-
phalloin (Formula 1c) und ^{14}C-N[ind]-methylphalloidin (Wieland and Govindan, 1974)(Formula 1d). Quite recently we obtained further evidence by
spectroscopy (Wieland et al., 1975b). Fig. 8 shows the difference spectrum between actin and phalloidin in two compartments of a tandem cu-

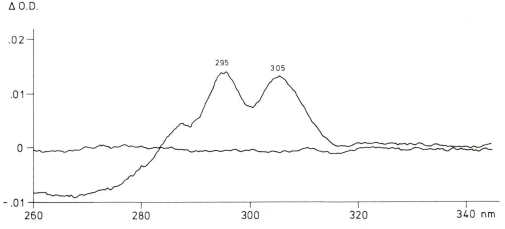

Fig. 8. Difference spectrum of the combination of actin and phalloidin,
each 1.09 × 10[-5] M

vette. Clearly, at 295 nm and 305 nm the u.v. spectrum is perturbed by the interaction of the toxin with actin. In control experiments phalloidin added to bovine serum albumin or the nontoxic secophalloidin (Formula 1h) together with actin produced only minimal deviations from the zero line. The changes in the region of 300 nm seen with phalloidin plus actin most probably originate from the indolylthioether moiety of the toxin, which evidently participates in the interaction.

The difference of absorptivity at 305 nm is linearly dependent on the phalloidin concentration below a ratio of toxin/actin ~ 0.5, the binding sites appear saturated at molar ratios around 1. This is in accordance with results obtained from measurements of the accelerating effect of phalloidin on the polymerization of G-actin and its stabilizing effect on F-actin referred to before (Figs. 3 and 7).

D. Phalloidin and Cytochalasin B

Cytochalasin B (CB), a metabolite of the fungus *Helminthosporium dematioideum* obviously weakens the F-actin structure. This can be concluded from the fact that CB inhibits cellular functions linked to actin-like microfilaments (for reviews see Wessels et al., 1971; Allison, 1973), and that it decreases the viscosity of F-actin solutions (Spudich and Lin Shin, 1972). Following our observation that CB prevented the phalloidin-induced formation of microfilaments in cell membrane preparations of rat liver (Löw et al., 1974), we inspected more closely the system F-actin-CB-phalloidin. We learned that phalloidin is able to antagonize the weakening effect which CB exerts on the F-actin structure (Löw et al., 1975). This was demonstrated by viscosimetry and by measuring the ATPase activity.

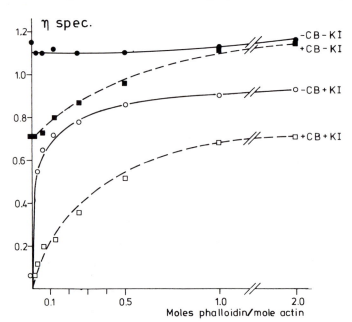

Fig. 9. Reversal by phalloidin (0-100 nmoles) of the viscosity reduction of 52 nmoles actin induced by KI and/or cytochalasin B (400 μmoles) in presence of 0.75 mM $MgCl_2$ without KCl

Fig. 9 shows the specific viscosity of F-actin solutions obtained by polymerization of G-actin in presence of Mg^{++}, ADP and phalloidin

(0-100 nmoles), K^+ being absent. After addition of KI the well known drop of viscosity was visible, which was counteracted by phalloidin in the way already shown in Fig. 5.

When the samples were incubated for 10 min at 35°C after addition of 400 nmoles CB three hours after phalloidin, a lower viscosity was found (Löw et al., 1975) particularly in presence of low concentrations of the toxin, which antagonized the disrupting effect of CB when added in sufficiently high concentrations (about 1 mole toxin pro 8 moles CB). When KI and CB were present together, more phalloidin was needed to reach a particular increased viscosity than with KI present alone. Also the level obtained with phalloidin + KI alone is higher than with KI plus CB. It should be mentioned that KCl apparently has an additional stabilizing effect, for the viscosity drop after CB, the so called "Spudich effect" could not be observed when KCl was also present, but the synergistic action of CB and KI against phalloidin was visible also in this system.

The weakening of the F-actin structure by CB is not only revealed by its influence on viscosity. From Fig. 10 it can be seen that F-actin in the presence of CB exhibits an ATPase activity of a similar rate as that induced by sonic vibration (Löw et al., 1975). The activity

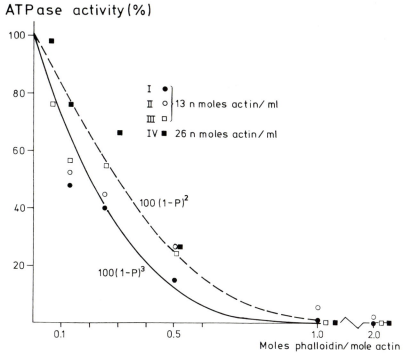

Fig. 10. ATPase activity of F-actin induced by cytochalasin B and its inhibition by phalloidin

is observable only when KCl is absent; as in the viscosity test, KCl also has a stabilizing effect on F-actin here. The ATPase activity observable in the absence of K^+, like that produced by sonic vibration, is not very high: the rate of liberation of inorganic phosphate is around O.2 mol P$_i$ per mole actin per minute. The curve in Fig. 10 shows that the ATPase decreases with increasing amounts of phalloidin.

As in the case of sonic vibration (Fig. 7) one can conclude that ATPase activity in the presence of CB is only possible when all actin subunits are free from phalloidin but that in order to inhibit the contribution of about 3 subunits to the ATPase activity only 1 of them needs to be combined with phalloidin.

E. Conclusions and Questions

What is the mechanism of the phalloidin action and why does stabiliza-tion of actin lead to impairment of hepatocyte membranes? The accelera-tion of G-actin polymerization could simply be explained by assuming that the protein, after having picked up a molecule of phalloidin, gains the ability to combine irreversibly with a second G-actin and that this process continues under formation of Ph-actin.

Still less is known about the relationship between phalloidin - actin interaction and impairment of hepatocyte membranes. "Cytoplasmic" (re-view by Pollard and Weihing, 1973) actin has been detected as a protein component of nearly all non-muscle cells not only of metazoa but also in protozoa, and has been made responsible, together with myosin, for movements like amoeboid locomotion, cytoplasmic streaming, cytokinesis, morphogenetic changes, secretion and resorption processes. The presence of actin (and myosin) has been demonstrated by indirect immunofluores-cence in cryostat sections of rat liver using human smooth muscle auto-antibody - which is an antiactin, present in the sera of most cases of acute infective hepatitis (Farrow et al., 1970) - and fluorescing anti-human IgM conjugate (Farrow et al., 1971). A polygonal pattern of stain-ing outlining parenchymal cells was observed, which points to the close proximity of the membrane or to the membrane itself as the locus of the cell actin. Together with the finding of myosin closely associated with cytoplasmic membranes (Willingham et al., 1974) (although of cultured cells) one could imagine that their interaction has something to do with a function of the membrane and that this interaction could be regulated by the polymerization state of actin. Not only phalloidin, which causes a freezing of the G-F-actin cycle in the F-form, causes an impairment of the hepatocyte membrane, but also CB has been stated by Jahn (1973) to cause vacuolization of the parenchyma cells of a perfused rat liver preparation, presumably as a consequence of its F-actin weakening effect. Therefore it seems conceivable that a poly-merization-depolymerization cycle of cell actin may be vital at least for liver cells.

References

Allison, A.C.: Ciba Foundation Symp. 14 (sew series) pp. 109-143. Am-sterdam-London-New York: Association of Scientific Publ. 1973.
Asakura, S.: Biochim. Biophys. Acta 52, 65-75 (1961).
Dancker, P., Löw, I., Hasselbach, W., Wieland, Th.: Biochim. Biophys. Acta 400, 407-414 (1975).
Farrow, L.J., Holborow, E.J., Brighton, W.D.: Nature New Biol. 232, 186-187 (1971).

Farrow, L.J., Holborow, E.J., Johnson, G.D., Lamb, S.G., Stewart, S.S.,
Taylor, P.E., Zuckermann, A.J.: Brit. Med. J. 2, 693 (1970).
Faulstich, H., Wieland, Th.: Eur. J. Biochem. 22, 79-86 (1971).
Faulstich, H., Wieland, Th., Jochum, Chr.: Liebigs Ann. Chem. 713, 186
(1968).
Frimmer, M., Gries, J., Hegner, D., Schnorr, B.: Naunyn-Schmiedebergs
Arch. Pharmacol. 258, 197-214 (1967).
Govindan, V.M., Faulstich, H., Wieland, Th., Agostini, B., Hasselbach,
W.: Naturwissenschaften 59, 521-522 (1972).
Govindan, V.M., Rohr, G., Wieland, Th., Agostini, B.: Hoppe-Seyler's
Z. Physiol. Chem. 354, 1159-1161 (1973).
Jahn, W.: Naunyn-Schmiedebergs Arch. Pharmacol. 275, 405-418 (1972).
Jahn, W.: Naunyn-Schmiedebergs Arch. Pharmacol. 278, 413-434 (1973).
Lazarides, E., Lindberg, U.: Proc. Nat. Acad. Sci. USA 71, 4742-4746
(1974).
Lengsfeld, A., Jahn, W.: Cytobiologie 9, 391-400 (1974).
Lengsfeld, A.M., Löw, I., Wieland, Th., Dancker, P., Hasselbach, W.:
Proc. Nat. Acad. Sci. USA 71, 2803-2807 (1974).
Löw, I., Dancker, P., Wieland, Th.: FEBS Letters 54, 263-265 (1975).
Löw, I., Lengsfeld, A.M., Wieland, Th.: Histochemistry 38, 253-258
(1974).
Löw, I., Wieland, Th.: FEBS Letters 44, 340-343 (1974).
Lutz, F., Glossmann, H., Frimmer, M.: Naunyn-Schmiedebergs Arch. Phar-
macol. 273, 341-351 (1972).
Pollard, T.D., Weihing, R.R.: Critical Reviews in Biochemistry 2,
1-65 (1973).
Schäfer, A., deVries, J.X., Faulstich, H., Wieland, Th.: FEBS Letters
57, 51-54 (1975).
Spudich, J.A., Lin Shin: Proc. Nat. Acad. Sci. USA 69, 442-446 (1972).
Wessels, N.K., Spooner, B.S., Ash, J.F., Bradley, M.O., Luduena, M.A.,
Taylor, E.L., Wrenn, J.T., Yamada, K.M.: Science 171, 135-143 (1971).
Wieland, Th.: Science 159, 946 (1968).
Wieland, Th., Govindan, V.M.: FEBS Letters 46, 351-353 (1974).
Wieland, Th., Jordan de Urries, M.P., Indest, H., Faulstich, H., Gieren,
A., Sturm, M., Hoppe, W.: Liebigs Ann. Chem. 1570-1579 (1974).
Wieland, Th., deVries, J.X., Schäfer, A., Faulstich, H.: manuscript
in preparation (1975a).
Wieland, Th., deVries, J.X., Schäfer, A., Faulstich, H.: FEBS Letters
54, 73-75 (1975b).
Wieland, Th., Wieland, O.: In: Microbial Toxins (eds. S. Kadis, A.
Ciegler, S.J. Ajl), p. 249-280. New York: Academic Press 1972.
Willingham, M.C., Ostlund, R.E., Pastan, I.: Proc. Nat. Acad. Sci.
USA 71, 4144-4148 (1974).

Discussion

Dr. Wilkie: Is anything known about the effect of phalloidotoxin on
the contraction of glycerinated muscle fibers? It would be most interesting to know whether or not the stabilization of actin alters the
contraction process.

Dr. Wieland: We only know until now that the binding of actin to myosin
is not inhibited by phalloidin. It has not yet been tested in a system
with sliding filaments.

Dr. Huxley: Does the evidence about production of filaments by phalloidin on isolated membranes indicate that the membranes themselves
initially contain actin in the G-form?

Dr. Wieland: That is the question, yes, the purification of membrane fractions does not lead to a 100% pure preparation. It appears to me that about 50-60% of this preparation, which is obtained by differential centrifugation, consists of fragments of membrane material, so we are not sure from where the actin comes. One suggestion is that it is a component of the membrane and the other one is that it is very closely associated with the membrane.

Dr. Gergely: Did I understand correctly that while KJ prevents polymerization of G-actin by phalloidin, Ph-F-actin is stable in the presence of KJ?

Dr. Wieland: Yes, it depends on the concentration of phalloidin. When the concentration of phalloidin is high enough then it will overcome the effect of every antagonist, but not more than 0.6 M KJ. In the presence of 0.6 M KJ no polymerization occurs with phalloidin. When the F-actin has been formed already and has reacted with phalloidin then the Ph-actin will be stable against even higher concentrations of KJ.

Dr. Fischer: When a foreign compound such as phalloidin reacts so specifically with a molecule such as actin, one always wonders whether this interaction is purely accidental or whether a site exists that would normally accept a given physiological compound. Surely one cannot expect actin to have evolved just to produce a phalloidin binding site. So my question is: do you know of any physiological compound that might react with that site and, in essence, compete with phalloidin?

Dr. Wieland: No, there is no such indication.

Dr. Calissano: Is there any receptor for phalloitoxins on live cells?

Dr. Wieland: I think G-actin is the receptor for phalloidin or also F-actin.

Dr. Calissano: What is the association constant of phalloidin for actin?

Dr. Wieland (added in proof): This is under investigation.

Final Comments

The Mosbach colloquia were held to fulfil two purposes: firstly, to review a field of main interest for non-specialists; secondly, to stimulate the exchange of new ideas and new concepts among specialists from different disciplines.

Recently, interest has been focused on muscle as an example of a motile system, which seems to be present in similar forms in many primitive cells. In the early days, research into contractile proteins was carried out by people who successfully combined physiology with physiological chemistry. This period is connected with the name of H.H. Weber. Later, the work on muscular proteins diversified into many areas of research, such as: electronmicroscopy, X-ray studies, kinetics, protein chemistry, enzymology etc. Now, for the first time it seems that this mosaic is being pieced together to form a picture which gives some idea of the organisation and function of a whole organ. In my opinion this convergence and mutual stimulation of different disciplines is best exemplified by the fact that one begins to see how an enzyme, phosphorylase kinase, originally found as a highly specialised, regulatory protein in carbohydrate metabolism, has probably evolved from a structural protein, actin. This correlation can even be further extended, in that this kinase contains an ATP-splitting catalytic centre combined with the structural protein, actin, in a similar fashion to the combination of the ATP-splitting enzyme myosin with actin.

This is just one point which I found very exciting and which I choose as an example of a new development, not to mention many other equally important findings which were reported at this meeting. However, this has not detracted at all from my love of enzymology which will always remain close to my heart.

My hope is that some younger students will also become as fascinated in this area as I myself, and that the ideas profoundly influenced by H.H. Weber, to whom we dedicated this Symposium, will continue to grow and also to work again in our country.

Finally, I should like to thank all those who helped organise and finance this meeting and who helped in the production of this progress report.

Professor Dr. L. Heilmeyer

Subject Index

Molecular Biology, Biochemistry and Biophysics

Editors: A. Kleinzeller, G. F. Springer, H. G. Wittmann

Springer-Verlag Berlin Heidelberg New York

Colloquien der Gesellschaft für Biologische Chemie in Mosbach/Baden

Inhibitors. Tools in Cell Research
20. Colloquium am 14.—16. April 1969
Editors: Th. Bücher, H. Sies

Mammalian Reproduction
21. Colloquium am 9.—11. April 1970
Editors: H. Gibian, E. J. Plotz

The Dynamic Structure of Cell Membranes
22. Colloquium am 15.—17. April 1971
Editors: D. F. Hölzl Wallach, H. Fischer

Protein-Protein Interactions
23. Colloquium am 13.—15. April 1972
Editors: R. Jaenicke, E. Helmreich

Regulation of Transcription and Translation in Eukaryotes
24. Colloquium am 26.—28. April 1973
Editors: E. K.-F. Bautz, P. Karlson, H. Kersten

Biochemistry of Sensory Functions
25. Colloquium am 25.—27. April 1974
Editor: L. Jaenicke